Studies in Fuzziness and Soft Computing, Volume 121

http://www.springer.de/cgi-bin/search_book.pl?series=2941

Editor-in-chief
Prof. Janusz Kacprzyk
Systems Research Institute
Polish Academy of Sciences
ul. Newelska 6
01-447 Warsaw
Poland
E-mail: kacprzyk@ibspan.waw.pl

Jonathan Lee (Ed.)

Software Engineering with Computational Intelligence

Spri

Heid
Ne
Hon
1

Jonathan Lee (Ed.)

Software Engineering with Computational Intellingence

Springer

Prof. Jonathan Lee
National Central University
Dept. of Computer Science
320 Chung-Li
Taiwan R.O.C
E-mail: yjlee@selab.csie.ncu.edu.tw

ISSN 1434-9922
ISBN 3-540-00472-6 Springer-Verlag Berlin Heidelberg New York

Library of Congress Cataloging-in-Publication-Data applied for

A catalog record for this book is available from the Library of Congress.

Bibliographic information published by Die Deutsche Bibliothek
Die Deutsche Bibliothek lists this publication in the Deutsche Nationalbibliographie; detailed
bibliographic data is available in the internet at <http://dnb.ddb.de>.

Springer-Verlag Berlin Heidelberg New York
a member of BertelsmannSpringer Science+Business Media GmbH
http://www.springer.de

© Springer-Verlag Berlin Heidelberg 2003
Printed in Germany

Typesetting: data delivered by editor
Cover design: E. Kirchner, Springer-Verlag, Heidelberg
Printed on acid free paper   62/3020/M - 5 4 3 2 1 0

Foreword

It is not an exaggeration to view Professor Lee's book, " Software Engineering with Computational Intelligence," or SECI for short, as a pioneering contribution to software engineering. Breaking with the tradition of treating uncertainty, imprecision, fuzziness and vagueness as issues of peripheral importance, SECI moves them much closer to the center of the stage. It is obvious, though still not widely accepted, that this is where these issues should be, since the real world is much too complex and much too ill-defined to lend itself to categorical analysis in the Cartesian spirit.

As its title suggests, SECI employs the machineries of computational intelligence (CI) and, more or less equivalently, soft computing (SC), to deal with the foundations and principal issues in software engineering. Basically, CI and SC are consortia of methodologies which collectively provide a body of concepts and techniques for conception, design, construction and utilization of intelligent systems. The principal constituents of CI and SC are fuzzy logic, neurocomputing, evolutionary computing, probabilistic computing, chaotic computing and machine learning. The leitmotif of CI and SC is that, in general, better performance can be achieved by employing the constituent methodologies of CI and SC in combination rather than in a stand-alone mode.

In what follows, I will take the liberty of focusing my attention on fuzzy logic and fuzzy set theory, and on their roles in software engineering. But first, a couple of points of semantics which are in need of clarification.

What are the differences between uncertainty, imprecision, fuzziness and vagueness? Avoiding a lengthy discussion, I will confine myself to clarifying the difference between fuzziness and vagueness-two distinct concepts which are frequently used in the same sense.

Before the advent of fuzzy set theory and fuzzy logic, the term that was used in the literatures of philosophy and logic to describe fuzziness was vagueness. But, in fact, in English at least, fuzziness and vagueness have distinct meanings. More specifically, vagueness relates to insufficient specificity whereas fuzziness relates to unsharpness of class boundaries. For example, the proposition "I will be back in a few minutes," is fuzzy but not vague, while "I will be back sometime," is both fuzzy and vague. In general, vagueness is a property of propositions but not predicates, while fuzziness applies to both propositions and predicates, in the sense that a proposition is fuzzy if it contains a fuzzy predicate.

Another point that is in need of clarification is the meaning of "fuzzy logic." Frequently, what is not recognized is that "fuzzy logic" has two distinct meanings. In a narrow sense, fuzzy logic is a logical system which deals with modes of reasoning which are approximate rather than exact. In a broader sense, which is in prevalent use today, fuzzy logic (FL) is not just a logical system but a collection of concepts and techniques for dialing with imprecisely defined classes and relations.

Thus, FL subsumes fuzzy logic in its narrow sense, and has a much broader scope. What is used in SECI is FL. What should be underscored is that in FL everything is-or is allowed to be-a matter of degree. It is this fundamental facet of FL that makes it possible to construct models-in every field, including software engineering-which are much closer to reality than those based on bivalent logic and classical analysis.

A case in point is the fuzzy-logic-based treatment in Chapter 3 of the important issue of requirement engineering. More specifically, in real world settings it is rarely the case that the goals and constraints are rigid, with no compromises allowed. Employment of FL makes it possible to define requirements and associated trade-offs though the use of concepts and techniques drawn from fuzzy logic and, especially, test-score semantics and fuzzy aggregation operators. Underlying these concepts and techniques is the fact that even when the ultimate goal is crisp, as it is in the case of chess, the intermediate goals are fuzzy because the relationship between the terminal goal and the intermediate goals is too complex to lend itself to precise analysis.

One of the principal tools in the armamentarium of fuzzy logic is the calculus of fuzzy if-then rules (CFR). This tool is employed with skill and insight in many chapters in SECI, especially in the context of fuzzy objects, fuzzy relations and fuzzy object-oriented languages and databases. It is of historical interest to note that my 1973 paper in which this calculus was outlined was greeted by many leading system theorists at the time with derision and contempt. A sample is the comment made by Kalman after hearing my first exposition in 1972: "I would like to comment briefly on Professor Zadeh's presentation. His proposals could be severely, ferociously, even brutally criticized from a technical point of view. This would be out of place here. But a blunt question remains: Is Professor Zadeh presenting important ideas or is he indulging in wishful thinking?" Today, thirty years later, history provides an answer. The calculus of fuzzy if-then rules and the underlying concept of the linguistic variable are employed in almost all applications of fuzzy logic in the realms of control, system analysis, knowledge representation, decision analysis and optimization.

A new and important development in fuzzy logic which may well have a substantial bearing on software engineering is the methodology of Computing with Words and Perceptions (CWP). A concept which plays a key role in CWP is that of Precisiated Natural Language (PNL). The language of fuzzy if-then rules is a sublanguage of PNL.

In CWP, the objects of computation are words and perceptions describes in a natural language. So far as PNL is concerned, it consists of propositions drawn from a natural language, NL, which are precisiable through translation into a precisiation language. In the case of PNL, the precisiation language is the language of generalized constraints (GCL). By construction, GCL is maximally expressive. To reason with perceptions, propositions in PNL are translated into what is referred to as Protoformal Language (PFL). A modular multiagent Deduction Module in PNL contains a large variety of protoformal rules of deduction which govern goal-directed generalized constraint propagation.

At this juncture, CWP and PNL are in initial stages of development. Eventually, I believe, they will prove to be of high relevance to software engineering, opening the door to new applications and new ideas for further advances.

In conclusion, Software Engineering with Computational Intelligence is a work whose importance cannot be overstated. SECI is a must reading for those who are involved, in one way or another, with the development or use of software engineering and its methodologies. Professor Lee, the contributors to SECI and the publisher, Springer, deserve our thanks and congratulations for producing a work that presents, with high authority and impressive expository skill, a wealth of information and new ideas in one of the most important domains of information science and technology.

Berkeley, CA,   *Lotfi A. Zadeh*
November 18, 2002

Contents

1 Introduction to Software Engineering with Computational Intelligence

Jonathan Lee

Department of Computer Science and Information Engineering
National Central University
Chungli, Taiwan
yjlee@selab.csie.ncu.edu.tw

1.1 Introduction

There are two distinct features that characterize software engineering: (1) it is an application of engineering to software, that is, the application of a systematic, disciplined, quantifiable approach to the development, operation, and maintenance of software; and (2) it is a transformation process, that is, the transformation that models the real world into its corresponding software world.

Therefore, one of the deepest traditions in software engineering is that of according respectability to what is quantitative, precise, rigorous, and categorically true. However, as Lotfi Zadeh has put it [19], the world we live is pervasively imprecise, uncertain, and hard to be categorical about. Moreover, the increasing demand for complex applications in diversified areas imposes a great challenge on developing software systems in order to deal with imprecise and uncertain information.

This book on software engineering with computational intelligence describes various treatments to software modeling and formal analysis that exploit the technologies developed in computational intelligence. These treatments enable the extension of computational intelligence to various phases in software life cycle along three dimensions:

- Managing fuzziness resided in the requirements, including the formulation of imprecise requirements using fuzzy Z and the trade-off analysis for conflicting requirements.
- Coping with fuzzy objects and imprecise knowledge, including the modeling of classes, rules, ranges of attributes, inclusion relationship, uncertain association, and fuzzy inference.
- Handling uncertainty encountered in quality prediction, including the utilization of neural networks, fuzzy neural networks, and fuzzy clustering, to develop general predication models for spotting the most trouble-prone modules early in the software life cycle.

1.2   Managing Imprecise and Conflicting Requirements

Requirements engineering is normally seen as the processes by which software requirements are identified and specified. A major challenge in requirements engineering is that the requirements to be captured are imprecise in nature and usually conflicting with each other [17]. Blazer et al. [1] have argued that informality is an inevitable (and ultimately desirable) feature of the specification process.

Chapter 2 by Chris Matthews and Paul A. Swatman suggests the use of fuzzy set theory together with Z as a formalism for modeling imprecise requirements. They extend Z syntax and structures to develop a toolkit which defines the set operators, measures and modifiers necessary for the manipulation of fuzzy sets and relations. The toolkit embraces (1) the generic definitions for the domain and range of a fuzzy relation as well as those for domain and range restriction, and anti-restriction, (2) the generic definition for the min-max and max-min composition operators for fuzzy relation, (3) the definition for the relational inverse of a fuzzy relation and an abbreviation for the identity relation in terms of a fuzzy relation, and (4) a series of laws which establishes an isomorphism between the extended notation and conventional Z when applied to crisp sets.

Requirements trade-off analysis technique introduced by Jonathan Lee *et al.* in Chapter 3 formulates imprecise requirements based on Zadeh's canonical form in test-score semantics and soft conditions. The trade-off among vague requirements is analyzed by identifying the relationship between requirements, which could be either conflicting, irrelevant, cooperative, counterbalance, or independent. Parameterized aggregation operations, *fuzzy and/or*, are selected to combine individual requirements. An extended hierarchical aggregation structure is proposed to establish a four-level requirements hierarchy to facilitate requirements and criticalities aggregation through the *fuzzy and/or*. A compromise overall requirement can be obtained through the aggregation of individual requirements based on the requirements hierarchy. The proposed approach provides a framework for formally analyzing and modeling conflicts between requirements, and for users to better understand relationships among their requirements.

1.3   Coping with Fuzzy Objects and Imprecise Knowledge

Rumbaugh and his colleagues [13] have argued that OOM is a way of thinking about problems using models organized around real world concepts which are usually expressed in natural languages. As Lotfi Zadeh pointed out in [18], it is evident that almost all concepts in or about natural languages are almost fuzzy in nature. Dubois *et al.* [4] and Lano [6] have further advocated that object classes with fuzzy memberships values are therefore a natural representation framework for real world concepts. Variations of fuzzy object-oriented modeling have been proposed[1], such as modeling fuzzy object-oriented systems [12], fuzzy objects [8], and fuzzy typing [9] etc.

Chapter 4 by Guy de Tré *et al.* delineates a new generalized object-oriented approach to data engineering. They propose formal framework for the definition of a generalized object-oriented data model for the modeling of both crisply and non-crisply described information. This approach is based on modeling objects from the real-world as natural as possible, thereby explicitly dealing with the possibly uncertain and/or fuzzy nature of the available information and data about these objects. A generalized type system is formally defined and level-2 fuzzy sets have been used to model the structural aspects of a generalized type. Three kinds of operators are distinguished: (1) extended operators, which have the same behavior as their crisp counterpart for "crisp" arguments and result in an "undefined" domain value if at least one of the arguments represents a non-crisp value; (2) generalized operators, which are obtained as a generalization of their crisp counterpart by applying Zadeh's extension principle two successive times; and (3) specific operators, which have been provided for the handling of level-2 fuzzy sets and their "inner-level" fuzzy sets.

Several ways in which fuzzy theory may be applied to handle imprecision in database include: object recognition [5], imprecise data management [2,3,14], fuzzy query [10,11], and semantic data model [15,16], which are usually achieved by:

- associating a fuzzy degree of membership with each tuple of a relation;
- making use fuzzy inference mechanism to query database, for example, fuzzy similarity, fuzzy rules, graph-based operator, and fuzzy association algebra; and
- managing imprecise information through mechanisms such as possibility distribution, linguistic fuzziness, fuzzy implication, fuzzy inclusion, fuzzy predicate, and fuzzy graph.

Chapters 5, 6, and 7 are devoted to the application of fuzzy object-oriented modeling to database systems. Chapter 5 by Gloria Bordogna and Sergio Chiesa depicts a representation of imperfect spatial information based on linguistic qualifiers. By adopting a fuzzy object oriented database approach, new data types for representing geographic entities characterized by either unsharp boundaries or indeterminate boundaries and relationships are defined. In this framework, vague and uncertain attribute values can be specified as fuzzy subsets of the attribute domain. These fuzzy subsets are interpreted as possibility distributions so that the degree of membership of each domain value expresses its degree of possibility to be the actual manifestation. They

---

[1] Refer [7] for a more detailed description of contemporary fuzzy object-oriented modeling approaches.

also describe the problem of defining a flexible query language for fuzzy and
indeterminate objects. Chapter 6 by Fernando Berzal *et al.* proposes an architecture of a fuzzy object-oriented database model built over an existing
classical database system. Besides, they also explain the different structures
proposed in this model to handle data imperfection at the different levels
it may appear, showing how to implement these structures using classical
object-oriented features. Chapter 7 by Ali Dogru and Adnan Yazici presents
a domain oriented approach to developing spatiotemporal software systems.
In the suggested domain model, a class diagram based representation of features is employed, supported by a domain terminology dictionary. For the
implementation stage support, the domain offers a set of components and
architectural patterns for composing an application.

In the area of knowledge-based systems, fuzzy logic provides particular
advantage to applications which require the inference of a process that is
poorly understood, difficult to model, containing unknown or unmodeled
dynamics, nonlinear, or time variant. Chapter 8 by Mong-Fong Horng *et
al.* proposes a modeling framework that integrates fuzzy logic and object-
oriented paradigm for the development of knowledge-based systems. In this
framework, each fuzzy linguistic object (FLO) represents a fuzzy linguistic
variable or a cluster of fuzzy linguistic variable, and the problem-solving activity is realized by message passing and inheritance among linguistic objects.
They also develop a software development tool, including domain knowledge
editor, knowledge-level simulator, object model editor, object simulator, and
C++ programming editor, based on the FLO-based framework. A class library supporting the automatic code generation of FLO-based systems is also
implemented.

To cope with the changeability of the relevance and objectivity values due
to newly generated knowledge or evolving requirements, Bedir Tekinerdoğan
and Mehmet Aksit propose a model and an approach for evaluating domain
knowledge using fuzzy logic techniques, called fuzzy knowledge source evaluator (FKSE) (see Chapter 9). The FKSE takes as input the relevance and the
objectivity (fuzzy) values of a knowledge source and computes the abstract
quality. The FKSE is validated with an experimental cases study in which a
set of knowledge sources need to be evaluated by a transaction domain expert
and a group of novice transaction designers, for two distinct problem. This
approach provides a more precise evaluation and can better cope with the
evolution of the knowledge sources and the corresponding problems that are
expressed as software requirements.

1.4 Handling Uncertainty for Software Quality Prediction

A software quality model that predicts a statistical measure of reliability
for each module enables developers and managers to focus resource on the

most fault-prone modules early in the software life cycle. In current software-reliability research, the concern is how to develop general prediction models, which can be broken down into three categories: neural networks, fuzzy neural networks, and fuzzy clustering.

- *Neural Networks:* Existing models typically rely on assumptions about the development environments, the nature of software failures, and the probability of individual failures occurring. All these assumptions must be made before a project begins, therefore, it nearly impossible to develop a universal model that will provide accurate predictions under all circumstances. Neural network models have a significant advantage over analytic models, though, because they require only history as input, and no assumptions. Neural networks tend to compute approximate answers within the scope of some ill-defined generalization of the training set data which can be used to identify fault-prone modules early in development, and thus reduce the risk of operational problems with those modules.
- *Fuzzy Neural Networks:* Neural networks have found numerous applications in solving classification and prediction problems. They are capable of learning from data and produce good results even dealing with noisy or incomplete data. However, an internal behavior of a neural network is hidden from the user. By contrast, fuzzy rules are suitable for expressing the heuristic knowledge about metrics and quality factors from experts. The behavior of such systems is quite transparent, and results obtained can easily be interpreted and understood. A fuzzy neural network represents a hybrid that combines advantages of both neural networks and fuzzy systems to provide a powerful tool for software quality prediction.
- *Fuzzy Clustering:* In the area of software development it would be of great benefit to predict early in the development process those components of the software system that are likely to have a high error rate or that need high development effort. The main problem is that classes of software components are often not well separated, and that there are no well defined boundaries. Fuzzy clustering technique can be used for constructing quality models to identify outlying software components that might cause potential quality problems.

In Chapter 10, Zhiwei Xu and Taghi M. Khoshgoftaar describe a new rule-based fuzzy classification modeling approach as a method for identifying fault-prone software modules, early in the software development life cycle. The modeling approach, based on software metrics and the proposed rules generation technique, is applied to extract fuzzy rules from numerical data. In addition, it also provides a convenient way to modify rules according to the costs of different misclassification errors.

In order to develop effective and useful methodologies, techniques and tools, a solid understanding of software development processes is needed. To ensure successful development, software engineers and developers shall understand all components, relations, rules and constrains related to software development. An interesting question is what techniques and methods should be used to extract such knowledge and gain understanding of software processes. One of the possible approaches is analysis of software data.

Chapter 11 by Marek Reformat and Witold Pedrycz proposes a general design methodology for granular modeling that is accuracy and transparency. This approach can be used to perform two tasks of processing software engineering data - prediction and association discovery. A model of a process of development effort of software modules has been built. The model has led to the recognition of significant relationships between software measures. Software engineers can use the model in two different ways: (1) to predict a number of needed changes in a module during development process; and (2) to improve their knowledge about relationship between the values of a selected set of software measures and development effort.

1.5   Conclusion

In summary, the confluence of software engineering and computation intelligence provides a powerful tool to help address the increasing demand for complex applications in diversified areas, the ever-increasing complexity and size of software systems, and the imperfect information inherited in nature.

The reader is invited to explore the following chapters for more detailed information about how technologies developed in computational intelligence are extended and applied to software engineering.

References

1. R. Balzer, N. Goldman, and D. Wile. Informality in program specifications. *IEEE Transactions on Software Engineering*, 4(2):94–103, 1978.
2. G. Bordogna, D. Lucarella, and G. Pasi. A fuzzy object oriented data model. In *Proceedings of the Third IEEE Conference on Fuzzy Systems*, pages 313–318, 1994.
3. G. Bordogna and G. Pasi. Graph-based interaction in a fuzzy object oriented database. *International Journal of Intelligent Systems*, 16(7):821–842, 2001.
4. D. Dubois, H. Prade, and J.P. Rossazza. Vagueness, typicality and uncertainty in class hierarchies. *International Journal of Intelligent Systems*, 6:161–183, 1991.
5. C. Granger. An application of possibility theory to object recognition. *Fuzzy Sets and Systems*, 28(3):351–362, 1988.
6. K. Lano. Combining object-oriented representations of knowledge with proximity to conceptual prototypes. In *Proceedings of Computer Systems and Software Engineering*, pages 442–446, 1992.
7. J. Lee, J.Y. Kuo, and N.L. Xue. A note on current approaches to extending fuzzy logic to object-oriented modeling. *International Journal of Intelligent Systems*, 16(7):807–820, 2001.
8. J. Lee, N.L. Xue, K.H. Hsu, and S.J. Yang. Modeling imprecise requirements with fuzzy objects. *Information Sciences*, 118:101–119, 1999.

9. N. Marin, O. Pons, and M.A. Vila. A strategy for adding fuzzy types to an object-oriented database system. *International Journal of Intelligent Systems*, 16(7):863–880, 2001.

10. S. Na and S. Park. A fuzzy association algebra based on a fuzzy object-oriented data model. In *Proceedings of 20th International Computer Software and Applications Conference (COMPSAC'96)*, pages 276–281, 1996.

11. S. Nepal, M.V. Ramakrishna, and J.A. Thom. A fuzzy object query language (foql) for image database. In *Proceedings of Six International Conference on Database Systems for Advanced Applications*, pages 117–124, 1999.

12. W. Pedrycz and Z.A. Sosnowski. Fuzzy object-oriented system design. *Fuzzy Sets and Systems*, 99:121–134, 1998.

13. J. Rumbaugh, M. Blaha, W. Premerlani, F. Eddy, and W. Lorensen. *Object-Oriented Modeling and Design*. Englewood Cliffs, NJ: Prentice Hall, 1991.

14. K. Tanaka, S. Kobayashi, and T. Sakanoue. Uncertainty management in object-oriented database system. *DEXA*, pages 251–256, 1991.

15. A. Yazici, R. George, and D. Aksoy. Design and implementation issues in the fuzzy object-oriented data model. *Information Sciences*, 108(1-4):241–260, 1998.

16. A. Yazici and M. Koyuncu. Fuzzy object-oriented database modeling coupled with fuzzy logic. *Fuzzy Sets and Systems*, 89:1–26, 1997.

17. J. Yen, X. Liu, and S.H. Teh. A fuzzy logic-based methodology for the acquisition and analysis of imprecise requirements. *Concurrent Engineering: Research and Applications*, 2:265–277, 1994.

18. L.A. Zadeh. Test-score semantics as a basis for a computational approach to the representation of meaning. *Literacy Linguistic Computing*, 1:24–35, 1986.

19. L.A. Zadeh. Soft computing and fuzzy logic. *IEEE Software*, 11(6):48–56, 1994.

2 Fuzzy Concepts and Formal Methods

Chris Matthews[1] and Paul A. Swatman[2]

[1] Department of Information Technology,
School of Business and Technology
La Trobe University, P.O. Box 199 Bendigo 3552, Victoria, Australia,
Phone: +61 3 54447350, Fax: +61 3 54447998
c.matthews@bendigo.latrobe.edu.au
[2] Stuttgart Institute of Management and Technology
Filderhauptstrße 142
70599 Stuttgart, Germany
and
School of Management Information Sciences,
Deakin University, Burwood, Victoria, Australia

2.1 Introduction

Requirements engineering or analysis is normally seen as the process by which software requirements are identified and specified, and has been described as a 'process of discovery, refinement, modelling and specification' [36]. System modelling is a useful tool in this process. We can create models to express our understanding of a current system problem, to identify areas for change or to describe the final product to be built. They permit the partitioning of complex system problems into smaller and more manageable units, and can be used as a basis for communication and validation with the client. Many of the modelling approaches use semi-formal diagramming techniques resulting, for example in data flow diagrams and entity relationship diagrams, and may be supported by explanatory text. However these approaches lack the well defined and formal syntax that permits precise analysis, and models expressed in these terms are open to possible misunderstanding and ambiguity. At some point we need to express the model more formally and to subject it to precise and rigorous analysis. The use of a formal specification language as a modelling tool provides a possible way to achieve this.

Formal methods are a set of tools that allow the development of a complete, precise and correct specification for system properties and behaviour. Although most commonly used in the specification of safety critical software, it has been argued that they can and should be applied to all stages of the systems development lifecycle including the specification of user requirements [45].

One commonly used specification language is Z. Z is a model-based formal specification language based on typed set theory and first order predicate calculus. The basic building block in any Z specification is the schema. This is a named grouping of components that can be referred to throughout the specification. Schemas provide structure for a specification and describe system operations and permissable system states. The schema calculus allows more complex components of a specification to be built from a set of previously defined and simpler schemas. Z is a powerful analytical tool that facilitates system understanding through the development of a series of unambiguous, verifiable mathematical models [15,33,35,41]. These models can be used to predict system behaviour and to identify errors prior to implementation. Various levels of abstraction are possible. These may vary from a statement of requirements (to be used as a basis for communication and validation) to a more detailed and concrete software design document. Unlike some of the more informal graphical methods, Z is not open to differing interpretations, but instead allows the designer to prove, through rigorous mathematical reasoning, the properties of a specification [37].

2.1.1   Problem Types and Uncertainty

There are some problem domains that may not naturally be understood in precise or crisp terms. L. A. Zadeh's concept of a humanistic system is one such example. He defines a humanistic system as one 'whose behaviour is strongly influenced by human judgement, perceptions or emotions' [49]. Economic, political, legal and educational systems are all examples of such systems and Zadeh suggests that a 'single individual and his thought processes may also be viewed as a humanistic system'. One characteristic of systems of this type is that human decision making and judgement may take place in a climate of uncertainty. In the development of systems that attempt to model human decision making it has been recognised that it is necessary to deal with uncertain or ill-defined knowledge [3,4,14]. We use terms such as vagueness, ambiguity, contradiction and imprecision when describing such systems. Our difficulty in precisely representing these softer system problems is due in part to the uncertainty inherent in the problem domain.

Uncertainty may arise in several ways. For example we may have difficulty classifying objects or concepts into one or more classes based on a series of attributes. The uncertainty as to which class an object belongs could be simply due to a lack of information. This type of uncertainty has been described as arising from *information deficiency* and can be resolved by extra attribute information [21]. However uncertainty can also arise from some natural imprecision in the problem domain itself. The classification into precise classes may not be possible — not because we do not have enough information — but instead, because the classes themselves are not naturally discrete. Categorical membership of this type has been sometimes defined as *fuzziness* and it has been suggested that it is, in part, a property of natural language arising from *linguistic imprecision* [21,52]. It is those problem domains exhibiting fuzziness that are of interest to us here.

Clearly such problem domains are difficult to model in precise terms. Concepts or objects, whether they be individuals, organisational units, opinions etc. may not be easily or naturally characterised into precise groupings.

Instead we may be more interested in the extent to which something resembles a type or in the relative ranking of something within a class or type rather than a precise description. Additionally many of these concepts may only be partially understood and difficult to define or precisely measure. In this environment system requirements themselves may be difficult to express or model precisely. For example, the requirement that the performance of a new network be *as efficient as possible* may be sufficient as an approximate statement of network performance, without having to precisely understand or define the performance benchmarks necessary to measure it, or to work out what is precisely meant by *as efficient as possible*. In fact it may be difficult or unrealistic to assign a precise meaning to statements of this type. Concepts like *profits are high, system performance is poor, share prices are low* etc. may be sufficient to express an approximate understanding shared by a group of people.

There is some support for this in the literature. Prototype theory [9,20,31] suggests that when classifying objects we do so by comparing an instance or example against what we consider to be prototypical or an exemplar of that class. Thus we may see an overlap, rather than a crisp boundary between categories. In other words the overlap is natural and cannot be expressed any more precisely. In the measurement of a person's attitudes or preferences, it has been argued that respondents should be allowed to indicate their *extent* of agreement rather than simply responding *yes* or *no* [44]. More generally it has been suggested that they should be also permitted to indicate the uncertainty or fuzziness in their response [12,13]. Some recent comparative experimental results supports this approach [5], again suggesting that precise categorisation may be inappropriate when dealing with a person's preferences, attitudes or requirements. The qualitative and imprecise nature of some system requirements themselves has been recognised. These requirements may only be stated approximately or expressed heuristically. The concept of a soft functional requirement — one where the degree to which the requirement precondition holds influences the degree to which that requirement is met — has been introduced [25]. This allows an elasticity to be introduced into the system model, which may more naturally reflect the competing requirements within a problem domain.

2.1.2   A Possible Extension for Z

This current work focusses on the possible use of formal methods, and Z in particular, as a modelling tool for system problems containing uncertainty that arises from some natural imprecision or vagueness in the problem domain. However Z requires that we are able to categorise objects and concepts into precise types as the specification is developed. This may present a problem if imprecision or fuzziness is inherent to the problem domain itself. There is a danger that we will lose part of what we are attempting to represent. Z also requires that permissable system states and operations be precisely expressed. Schema pre-conditions and post-conditions either hold or they don't, a system operation either succeeds or fails — the idea that a requirement need only be *partially* met or that a given system state may only *approximately* hold is difficult to imagine, at least within the context of a Z specification. As suggested earlier, there are advantages in introducing a formal approach as early as possible during requirements determination. It may be useful if we are able to express those parts of the problem domain that appear imprecise or perhaps are only partially understood, in the formal model. Then we could refine the model as these issues are clarified. One could imagine both precise and imprecise concepts from the problem domain being expressed in the same formal model. This would be particularly useful if we attempt to develop formal models early in requirements determination, while elicitation is still occuring. In order to facilitate this approach to requirements engineering the specification language should allow us to both syntactically and semantically capture and represent imprecision and/or approximation. Fuzzy set theory (and fuzzy logic) offers such a possibility, particularly when dealing with naturally occuring imprecision.

2.1.3 Fuzzy Sets

Fuzzy set theory and fuzzy logic provides a mathematical basis for representing and reasoning with knowledge in uncertain and imprecise problem domains. Unlike boolean set theory where set membership is crisp (i.e. an element is either a member or it isn't), the underlying principle in fuzzy set theory is that an element is permitted to exhibit partial membership in a set. Fuzzy set theory allows us to represent the imprecise concepts (eg *motivated employees*, *high profits* and *productive workers*) which may be important in a problem domain within an organizational context. The common set operators such as negation, union and intersection all have their fuzzy equivalents and measures for fuzzy subsetness and fuzzy set entropy have been proposed [21,22,27]. Fuzzy logic deals with degrees of truth and provides a conceptual framework for approximate rather than exact reasoning. The truth of propositions such as a few *employees are motivated* or *productive workers lead to high profits* can be estimated and reasoned with [50,51].

Fuzzy set theory and fuzzy logic have been successfully applied to the development of industrial control systems [43] and commercial expert systems [10]. Fuzzy set theory, and related theories such as possibility theory [8], have been suggested as appropriate analytical tools in the Social Sciences [39,40]. The idea that over-precision in measurement instruments may present a methodological problem in psychological measurement has led to developments such as a fuzzy graphic rating scale for the measurement of occupational preference [13], fuzzy set based response categories for marketing applications [44] or a fuzzy set importance rating scheme for personnel selection [1]. Fuzzy set theory and fuzzy logic have also been used to model group decision making, particularly when the preference for one set of options over another is not clear cut [17]. The use of fuzzy propositions to capture the elasticity of 'soft' functional requirements has been proposed as a technique for modelling imprecision during requirements engineering for knowledge-based system development [25]. Research has also indicated that there may be some compatibility between fuzzy set theory and the meanings that we as humans place on the linguistic terms that are normally used to describe such sets [32,38,53,54]. This suggests that modelling techniques based on fuzzy set theory may lead to models that are closer to our cognitive processes and models than those based on boolean set theory.

2.1.4   Motivation

It is clear that fuzzy methods are useful as both an analytical and a descriptive tool in problem domains that are characterised by uncertainty and fuzziness. The motivation for the current research can therefore be summarised as follows.

- Given that there are some system problems, particularly those drawn from a humanistic context, that are not naturally modelled or understood in precise or crisp terms and given that we wish to retain the benefits of a specification language such as Z as a method for communication and validation, is it possible to build into the existing syntax the necessary semantics to capture the uncertainty, imprecision or vagueness characteristic of such systems?
- Fuzzy set theory is an established technique for representing uncertainty and imprecision and can be seen as a generalisation of boolean or crisp set theory. Given that Z is a set based specification language then it should be possible to provide a notation that incorporates fuzzy set ideas within the language itself while at the same time retaining the precision of any Z model.

This current research is concerned with the development of a suitable fuzzy set notation within the existing Z syntax. It is assumed that the existing schema calculus and logical structures of Z remain. A toolkit has been developed which defines the set operators, measures and modifiers necessary for the manipulation of fuzzy sets. The version of the toolkit summarised here is a modification to that previously presented [29,30][1]. It contains generic definitions for the domain and range of a fuzzy relation as well as those for domain and range restriction, and anti- restriction. Generic definitions for the *min-max* and *max-min* composition operators for fuzzy relations have been developed. The relational inverse of a fuzzy relation has been defined

---

[1] The reader is also directed to the *fuzzy logic toolkit* archive (accessed from www.ironbark.bendigo.latrobe.edu/staff/chrism/chrishp.html), where the most recent versions of the toolkit can be found.

and an abbreviation for the identity relation in terms of a fuzzy relation has been provided. A series of laws which establish an isomorphism between the extended notation presented here and conventional Z when applied to crisp sets (i.e. sets where the membership values are constrained to 0 or 1) has also been provided. The toolkit also identifies (and provides proofs for) the relevant laws from [41] that hold when partial set membership is permitted. In this chapter a summary of the toolkit is presented, including some sample laws. The reader is referred to [29] for the current version of the complete toolkit and proofs for the laws presented here.

The chapter is organised as follows: Section 2.2 introduces the fuzzy set repesentation scheme used in the toolkit and discusses some alternatives. Section 2.3 presents a summary of the toolkit itself. Section 2.4 introduces some examples intended to illustrate the use of the toolkit when modelling system problems of interest and section 2.5 develops some general guidelines for toolkit usage. The chapter concludes with some directions for ongoing work.

2.2   Fuzzy Set Representation in Z

A fuzzy set, μ, can be represented as a mapping from a reference set, X, to the real number interval $[0, 1]$.

$$\mu : X \to [0, 1]$$

The membership of each element of the reference set is given by $\mu(x)$, where $x \in X$. A fuzzy set can be imagined as a set of ordered pairs. Crisp sets are those where the membership values, $\mu(x)$, are constrained to either 1 or 0 for all $x \in X$. Sets where the set membership of each element is shown explicitly is referred to as being written in an *extended* notation. Crisp sets can be expressed in either the extended notation or in more conventional terms. For example, the set $A = \{(x_1, 1), (x_2, 1), (x_3, 0), (x_4, 0), (x_5, 1)\}$ could be written simply as $\{x_1, x_2, x_5\}$.

Z allows the definition of relations and functions between sets, and provides the necessary operations to manipulate them. If the reference set is considered to be a basic type within a Z specification then a fuzzy set can be defined as a function from the reference set to the interval $[0,1]$. However Z does not define the fuzzy set operators for union, intersection and set difference, or the fuzzy set modifiers such as *not*, *somewhat* and *very* which are needed for fuzzy set and fuzzy set membership manipulation. The toolkit provides the generic definitions, axiomatic descriptions and abbreviations for these operators, modifiers and measures. The T-norm and T-conorm operators $min\{\mu_1(x), \mu_2(x)\}$ and $max\{\mu_1(x), \mu_2(x)\}$ have been used to determine the membership of a reference set element x in the intersection and union of the two fuzzy sets, μ_1 and μ_2. These operators are well established and preserve many of the properties of their boolean equivalents [2,19,24]. When applied to crisp sets, *min* and *max* behave in the same way as do the existing operators, ∩, ∪, for boolean sets.

It is possible to visualise a fuzzy set other than as a function. A set-based representation scheme based on a set of nested subsets formed from the $\alpha cuts$ of a fuzzy set is a possibility [24,47]. An αcut is set of reference set elements whose membership in the fuzzy set is greater than or equal to some threshold value. For example the 0.5 αcut contains reference set elements whose membership in the fuzzy set is greater or equal to 0.5 A fuzzy set could be modelled as a set of such sets. Membership of a reference set element in one of these sets would indicate that the membership of the element in the fuzzy set is equal to or exceeds the membership value represented by the αcut. However the definition of an αcut relies on the ability to determine the membership of the reference element in the fuzzy set in the first place i.e. on the existence of some characteristic function that delivers set membership. This would suggest that a functional approach is more appropriate as the basic repesentation scheme for a fuzzy set and that a set-based representation could be induced through the concept of $\alpha cuts$ when required. The toolkit (summarised in section 2.3) provides a generic definition for the formation of an αcut for a fuzzy set.

Another alternative is to use a vector notation to provide a geometric rather than algebraic representation for a fuzzy set[22,23]. For a finite reference set, X, containing n elements we could define a series of fuzzy subsets. Each could be represented each an n-dimensional vector, where the vector components are the membership values of the corresponding elements. Furthermore we could visualise the fuzzy subset geometrically as a point in an n-dimensional space (or hypercube). Crisp subsets are represented by those points at the vertices of the n-dimensional hypercube. The closer a point is to the centre of the hypercube, the *more* fuzzy is the set it represents. The point at the centre of the hypercube is the one where all vector components are equal to 0.5. The set of possible points represents the *power set* of X (i.e. the set of all subsets of X — fuzzy and crisp).

This notation is useful as it provides an elegant visualisation of fuzzy set measures such as *cardinality, fuzzy entropy* and *degree of subsetness*.The notation is also simple as only the membership values are represented and manipulated — the reference set elements are implied by the ordering of the vector components. However its usefulness as a representation scheme within a set based, algebraic language such as Z appears limited. A vector notation is not a basic mathematical construct within Z, and would need to be defined. The notation does not explicitly define a mapping function from the reference set to the membership interval [0,1]. The membership vector for a particular fuzzy set would need to be enumerated rather than evaluated. Within a specification it may be necessary to evaluate the set membership of a particular reference set element. This is easily done if the membership function is available and explicitly stated. The alternative is to enumerate the membership value of each reference set element. For a large reference set this would become tedious and for infinite reference sets such as $\mathbb{R}$, not possible.

2.3   The Toolkit Summary

Two general principles have guided the preparation of the toolkit.

- Where applicable as much 'fuzziness' as possible is permitted. For example, rather than defining generalised union and intersection in terms of a crisp set of fuzzy sets, they are defined in terms of a *fuzzy* set of fuzzy sets. The domain and range restriction (and anti-restriction) for a fuzzy relation by a *fuzzy* rather than a crisp set is permitted and so on.

- When applied to crisp sets in the extended notation the toolkit definitions, abbreviations and descriptions must be isomorphic with conventional Z. Two functions, $\mathcal{P}$ (the extended set function) which maps sets of type $\mathbb{P}\,T$ in the conventional notation to a fuzzy set of type $\mathcal{F}T$ and $\mathcal{Q}$ (the core set function) which maps those reference set elements having full membership in a fuzzy set to a power set in the conventional notation, are included and are used in the proofs which attempt to establish this isomorphism.

The Z draft standard provides a given set, $\mathbb{A}$ which can be used to specify number systems [33,42]. It is assumed that the set of real numbers, $\mathbb{R}$ is defined as a subset of $\mathbb{A}$. It is also assumed that a division operator $/$ and the functions *sqrt*, *abs*, *min* and *max* have been defined for the set of real numbers. Furthermore it is assumed that the reader has a rudimentary understanding of the Z notation. If not the reader is directed to the Z archive[2] or the reference list (in particular [15] and [35]) for a basic overview of the language.

The toolkit has been type checked using the type checking software, ZTC[3] and can be used together with the default mathematical toolkit to type check specifications written in the extended notation.

2.3.1   Some basic definitions

Set membership is measured using the real number interval $[0,1]$.

$$\mathcal{M} == \{r : \mathbb{R} \mid 0 \le r \le 1\}$$

The generic symbol, $\mathcal{F}$ defines a fuzzy set as a function from a reference set(type) to the real number interval, $\mathcal{M}$.

---

[2] Accessed from www.afm.sbu.ac.uk/z/
[3] ZTC: A Type Checker for Z Notation, Version 2.01, May 1995 (Xiaoping Jia, Division of Software Engineering, School of Computer Science, Telecommunication, and Information Sciences, DePaul University, Chicago, Illinois, USA).

$$\mathcal{F}T == T \nrightarrow \mathcal{M}$$

The generic symbol, $\mathcal{C}$ constrains the membership values of elements in a fuzzy set to $\{0,1\}$. Sets of this type are crisp sets written in the extended notation.

$$\mathcal{C}T == T \nrightarrow \{0,1\}$$

The *extended set* function, $\mathcal{P}$ is used to map a power set of type T (in the conventional notation) to a fuzzy set, $\mathcal{F}T$ in the extended notation.

$$
\begin{array}{l}
[T] \\
\hline
\mathcal{P} : (\mathbb{P}\, T) \to (\mathcal{C}\, T) \\
\hline
\forall t : T;\ set : \mathbb{P}\, T\ \bullet \\
\qquad (t \in set \Leftrightarrow t \in \mathrm{dom}(\mathcal{P}(set)) \land \mathcal{P}(set)(t) = 1)\ \land \\
\qquad (t \notin set \Leftrightarrow t \in \mathrm{dom}(\mathcal{P}(set)) \land \mathcal{P}(set)(t) = 0)
\end{array}
$$

The *core set* function, $\mathcal{Q}$ is used to map those reference set elements having a membership of 1 in a fuzzy set of type $\mathcal{F}T$ to a power set of type T (in the conventional notation). This is sometimes referred to as the *core* of a fuzzy set [16].

$$
\begin{array}{l}
[T] \\
\hline
\mathcal{Q} : (\mathcal{F}T) \to (\mathbb{P}\, T) \\
\hline
\forall\, fun : \mathcal{F}T \bullet (\forall t : \mathrm{dom}\, fun\ \bullet \\
\qquad t \in \mathcal{Q}(fun) \Leftrightarrow fun(t) = 1 \\
\qquad t \notin \mathcal{Q}(fun) \Leftrightarrow fun(t) < 1)
\end{array}
$$

The toolkit also defines the *support set*, $\mathcal{S}$, of a fuzzy set in a similar fashion. The *support set* is the set of those reference set elements that exhibit some membership in the fuzzy set [16,26]. $\mathcal{S}$ is defined as a function mapping a set of type $\mathcal{F}T$ to that of type $\mathbb{P}\, T$. For sets $fun1, fun2$ of type $\mathcal{C}T$ the following holds.

$$\mathcal{Q}(fun1) = \mathcal{S}(fun1) \tag{2.1}$$

The αcut of a fuzzy set is the set of reference set elements whose membership values are greater or equal to α, where $0 \leq \alpha \leq 1$. A *strict* αcut is one where the membership values are greater than α. The notation used here (i.e. $[\_]_\alpha$ and $[\_]_{\underline{\alpha}}$) is taken from [24]. As indicated earlier the concept of an αcut is useful for a set-based rather than functional representation of a fuzzy set [47].

$$
\begin{array}{|l}
\hline
[T] \\
\hline
\left[\,_-\,\right]_{\alpha} : \mathcal{F}T \times \mathcal{M} \to \mathbb{P}\,T \\
\left[\,_-\,\right]_{\underline{\alpha}} : \mathcal{F}T \times \mathcal{M} \to \mathbb{P}\,T \\
\hline
\forall \alpha : \mathcal{M};\ fun : \mathcal{F}T;\ t : T \bullet \\
\quad t \in \left[\,fun\,\right]_{\alpha} \Leftrightarrow fun(t) \geq \alpha \wedge t \in \left[\,fun\,\right]_{\underline{\alpha}} \Leftrightarrow fun(t) > \alpha \\
\hline
\end{array}
$$

The *extended support set*, $\mathcal{ES}$ of a fuzzy set is a function of type $T \nrightarrow \mathcal{M}$ containing only that part of the fuzzy set where set membership is greater than zero.

$$
\begin{array}{|l}
\hline
[T] \\
\hline
\mathcal{ES} : (\mathcal{F}T) \to (\mathcal{F}T) \\
\hline
\forall fun : \mathcal{F}T \bullet \mathcal{ES}(fun) = fun \rhd \{0\} \\
\hline
\end{array}
$$

The toolkit also defines a function, total , which transforms a fuzzy set to one of type $T \to \mathcal{M}$. It is used in several of the definitions that follow to ensure that set membership is defined for all reference set elements.

$$
\begin{array}{|l}
\hline
[T] \\
\hline
\mathsf{total}\ : (\mathcal{F}T) \to (T \to \mathcal{M}) \\
\hline
\forall fun : (\mathcal{F}T);\ t : T \bullet \\
\quad\quad t \in \operatorname{dom} fun \Leftrightarrow (\mathsf{total}\ fun)(t) = fun(t) \\
\quad\quad t \notin \operatorname{dom} fun \Leftrightarrow (\mathsf{total}\ fun)(t) = 0 \\
\hline
\end{array}
$$

A *non-empty fuzzy set* is one where at least one reference set element has a membership greater than zero.

$$
\mathcal{F}_1 T == \{F : \mathcal{F}T \mid \exists\, t : T \bullet F(t) > 0\}
$$

A *finite fuzzy set* is one that has a finite number of elements with membership greater than zero.

$$
\mathsf{finite}\ \mathcal{F}T == \{F : \mathcal{F}T \mid F \rhd \{0\} \in \mathbb{F}(T \times \mathcal{M})\}
$$

A *non empty finite fuzzy set*, finite $\mathcal{F}_1$, is also defined in the toolkit. It is one where there is at least one element with a membership value greater than zero.

The following is a definition for the total membership relation for a fuzzy set. A zero membership relation, notin , is also provided and is defined in a similar way using $\{0\}$ as the range restricting set.

$$
\begin{array}{|l}
\hline
[T] \\
\hline
_-\ \mathsf{in}\ _- : T \leftrightarrow (\mathcal{F}T) \\
\hline
\forall t : T;\ fun : \mathcal{F}T \bullet t\ \mathsf{in}\ fun \Leftrightarrow t \in \operatorname{dom}(fun \rhd \{1\}) \\
\hline
\end{array}
$$

2.3.2   Some set measures

An empty set, empty , of type $\mathcal{F}T$ is one where all membership values are zero. The empty set is a finite fuzzy set.

$$
\begin{array}{|l}
\models [T] \\\hline
\quad \mathsf{empty} \ : \mathcal{F}T \\\hline
\quad \forall\, t : T \bullet \mathsf{empty}\ (t) = 0 \\
\end{array}
$$

A universal set, $\mathcal{U}$, of type $\mathcal{F}T$ is defined in the toolkit as one where all membership values are one.

The *cardinality* of a fuzzy set is defined as the sum of the membership values [23]. It only has meaning for a finite fuzzy set. *Counter* sums the membership values within sets of type $T \nrightarrow \mathcal{M}$ and count restricts the summation to membership values greater than zero in a finite fuzzy set.

$$
\begin{array}{|l}
\models [T] \\\hline
\quad \mathsf{counter}\ : (T \nrightarrow \mathcal{M}) \to \mathbb{R} \\\hline
\quad \forall\, fun : (T \nrightarrow \mathcal{M}) \bullet (\forall\, t : \mathrm{dom}\, fun \bullet \\
\qquad\qquad \mathsf{counter}\ (fun) = fun(t) + \mathsf{counter}\ (\{t\} \lhd fun)) \ \wedge \\
\qquad\qquad (fun = \varnothing) \Rightarrow (\mathsf{counter}\ (fun) = 0) \\
\end{array}
$$

$$
\begin{array}{|l}
\models [T] \\\hline
\quad \mathsf{count}\ : (\mathsf{finite}\ \mathcal{F}T) \to \mathbb{R} \\\hline
\quad \forall\, fun : \mathsf{finite}\ \mathcal{F}T \bullet \\
\qquad\qquad \mathsf{count}\ (fun) = \mathsf{counter}\ (\mathcal{ES}(fun)) \\
\end{array}
$$

2.3.3   Some set operators

Set *union* (or) and *intersection* (and) for fuzzy sets.

$$
\begin{array}{|l}
\models [T] \\\hline
\quad \_ \ \mathsf{or} \ \_ : (\mathcal{F}T) \times (\mathcal{F}T) \to (\mathcal{F}T) \\\hline
\quad \forall\, fun1, fun2 : \mathcal{F}T \bullet \\
\qquad\qquad \mathrm{dom}(fun1 \ \mathsf{or} \ fun2) = \mathrm{dom}\, fun1 \cup \mathrm{dom}\, fun2 \ \wedge \\
\qquad\qquad \forall\, t : \mathrm{dom}(fun1 \ \mathsf{or} \ fun2) \bullet \\
\qquad\qquad (fun1 \ \mathsf{or} \ fun2)(t) = max\{(\mathsf{total}\ fun1)(t), (\mathsf{total}\ fun2)(t)\} \\
\end{array}
$$

$$\begin{array}{|l}
\hline
\rule{0pt}{0pt}[T]\\
\hline
\_ \text{ and } \_ : (\mathcal{F}T) \times (\mathcal{F}T) \to (\mathcal{F}T)\\
\hline
\forall\, fun1, fun2 : \mathcal{F}T \bullet\\
\qquad \mathrm{dom}(fun1 \text{ and } fun2) = \mathrm{dom}\, fun1 \cap \mathrm{dom}\, fun2 \;\wedge\\
\qquad \forall\, t : \mathrm{dom}(fun1 \text{ and } fun2) \bullet\\
\qquad (fun1 \text{ and } fun2)(t) = min\{fun1(t), fun2(t)\}\\
\hline
\end{array}$$

The *complement* of a fuzzy set.

$$\begin{array}{|l}
\hline
\rule{0pt}{0pt}[T]\\
\hline
\text{not } : (\mathcal{F}T) \to (\mathcal{F}T)\\
\hline
\forall\, fun : \mathcal{F}T \bullet\\
\qquad \mathrm{dom}(\text{not } fun) = \mathrm{dom}\, fun \;\wedge\\
\qquad \forall\, t : \mathrm{dom}\, fun \bullet (\text{not } fun)(t) = 1 - fun(t)\\
\hline
\end{array}$$

Set *difference* between any two fuzzy sets.

$$\begin{array}{|l}
\hline
\rule{0pt}{0pt}[T]\\
\hline
\_ \overset{\mathcal{F}}{\setminus} \_ : (\mathcal{F}T) \times (\mathcal{F}T) \to (\mathcal{F}T)\\
\hline
\forall\, fun1, fun2 : \mathcal{F}T \bullet\\
\qquad fun1 \overset{\mathcal{F}}{\setminus} fun2 = fun1 \text{ and } (\text{not } fun2)\\
\hline
\end{array}$$

These operators have the same meaning for crisp sets written in either the conventional or extended notation. $\forall\, fun1, fun2 : \mathcal{C}T$,

$$\mathcal{Q}(fun1 \text{ and } fun2) = \mathcal{Q}fun1 \cap \mathcal{Q}fun2 \qquad (2.2)$$

$$\mathcal{Q}(fun1 \text{ or } fun2) = \mathcal{Q}fun1 \cup \mathcal{Q}fun2 \qquad (2.3)$$

$$\mathcal{Q}(fun1 \overset{\mathcal{F}}{\setminus} fun2) = \mathcal{Q}fun1 \setminus \mathcal{Q}fun2 \qquad (2.4)$$

The *min* and *max* operators are used for fuzzy set intersection and union to provide on the one hand, a generalisation of boolean set theory, and on the other, to preserve as much of the existing mathematical structure as possible. They are *commutative*, *associative*, *idempotent* and *distributive* [4]. When using $1 - fun(t)$ as the membership of the reference set element t in the complement

---

[4] Alternative operator definitions such as *product*, *bounded sum* and *mean* have received some intuitive and practical support [6,39,54]. However they sacrifice some of these properties in the fuzzy case. For example the product operators (i.e. $fun1(t) * fun2(t)$ for intersection and $fun1(t) + fun2(t) - fun1(t) * fun2(t)$ for union) are neither idempotent or distributive for partial set membership.

of the fuzzy set *fun*, it has been shown that De Morgan's Laws hold for sets
of type $T \to \mathcal{M}$ [2,26] i.e.

$$\text{not } (fun1 \text{ and } fun2) = \text{not } (fun1) \text{ or not } (fun2) \tag{2.5}$$

$$\text{not } (fun1 \text{ or } fun2) = \text{not } (fun2) \text{ and not } (fun2) \tag{2.6}$$

Only the *law of the excluded middle* is not valid in the fuzzy case. It only
holds for sets of type $\mathcal{C}\,T$.

$$fun \text{ and not } (fun) = \text{ empty} \tag{2.7}$$

This is expected and reflects the overlap between a fuzzy set and its complement.

The concept of a *generalised union*, $\overset{\mathcal{F}}{\bigcup}$, and that of a generalised *intersection*,
$\overset{\mathcal{F}}{\bigcap}$, for a fuzzy set of fuzzy sets is also defined in the toolkit. Membership of a
fuzzy set *fun* in a fuzzy set of fuzzy sets, A, could be interpreted as indicating
the *degree* to which *fun* will take part in the union or the intersection. A
constraint is placed on the definition of generalised intersection to ensure that
it is only formed from those sets that exhibit some membership in the fuzzy
set of fuzzy sets. When applied to a crisp set of fuzzy sets, generalised union
and intersection have the same meaning as fuzzy set union and intersection
(ie and and or).

2.3.4   Fuzziness, set equality and set inclusion

The definition for the *degree of fuzziness* of a finite fuzzy set is based on the
concept of fuzzy set entropy [22,23]. The degree of fuzziness of a fuzzy set can
be estimated by determining the degree of resemblence between the set and
the complement. This definition only relates to those reference set elements
that exhibit some membership in the fuzzy set. For a non-empty finite crisp
set, the membership value of each element in the extended support set can
only be one. The degree of fuzziness of such a set is zero.

$=[T]$ ==
  fuzzyEntropy : (finite $\mathcal{F}T$) $\to \mathbb{R}$

$\forall fun : \text{finite } \mathcal{F}T \bullet$
$fun = \text{ empty } \Rightarrow \text{ fuzzyEntropy}(fun) = 0 \land$
$fun \neq \text{ empty } \Rightarrow \text{ fuzzyEntropy}(fun) =$
counter $(\mathcal{ES}(fun) \text{ and not } (\mathcal{ES}(fun)))/\text{counter } (\mathcal{ES}(fun) \text{ or not } (\mathcal{ES}(fun)))$

A measurement of the *degree of equality* of two fuzzy sets.

$$\boxed{\begin{array}{l} [T] \\ \hline \_ \approx \_ : (\text{finite } \mathcal{F}T) \times (\text{finite } \mathcal{F}T) \to \mathcal{M} \\ \hline \forall \mathit{fun}1, \mathit{fun}2 : \text{finite } \mathcal{F}T \ \bullet \ (\mathit{fun}1 = \ \text{empty} \ \wedge \mathit{fun}2 = \ \text{empty} \ ) \Rightarrow \\ (\mathit{fun}1 \approx \mathit{fun}2 = 1) \ \wedge \\ (\mathit{fun}1 \neq \ \text{empty} \ \vee \mathit{fun}2 \neq \ \text{empty} \ ) \Rightarrow \\ \mathit{fun}1 \approx \mathit{fun}2 = \text{count} \ (\mathit{fun}1 \ \text{and} \ \mathit{fun}2)/\text{count} \ (\mathit{fun}1 \ \text{or} \ \mathit{fun}2)) \end{array}}$$

The expression, count $(\mathit{fun}1$ or $\mathit{fun}2)$ can only be zero when both $\mathit{fun1}$ and $\mathit{fun2}$ are the empty fuzzy set (i.e. where all membership values of the reference set are zero). In this case $\mathit{fun}1 = \mathit{fun}2$ and therefore the degree to which $\mathit{fun}1$ equals $\mathit{fun}2$ is one.

The following definitions for the *subsetness* relations for sets of type $\mathcal{F}T$ follow the original definitions [48] and define complete inclusion.

$$\boxed{\begin{array}{l} [T] \\ \hline \_ \overset{\mathcal{F}}{\subseteq} \_ : (\mathcal{F}T) \leftrightarrow (\mathcal{F}T) \\ \_ \overset{\mathcal{F}}{\subset} \_ : (\mathcal{F}T) \leftrightarrow (\mathcal{F}T) \\ \hline \forall \mathit{fun}1, \mathit{fun}2 : \mathcal{F}T \ \bullet \ (\mathit{fun}1 \overset{\mathcal{F}}{\subseteq} \mathit{fun}2) \Leftrightarrow \\ \qquad (\forall t : T \ \bullet \ (\text{total } \mathit{fun}1)(t) \le (\text{total } \mathit{fun}2)(t)) \ \wedge \\ \qquad (\mathit{fun}1 \overset{\mathcal{F}}{\subset} \mathit{fun}2) \Leftrightarrow (\mathit{fun}1 \overset{\mathcal{F}}{\subseteq} \mathit{fun}2 \wedge \mathit{fun}1 \neq \mathit{fun}2) \end{array}}$$

Subsetness has the same meaning for crisp sets written in either the conventional or extended notation. $\forall \mathit{fun}1, \mathit{fun}2 : \mathcal{C}T$,

$$\mathit{fun}1 \overset{\mathcal{F}}{\subseteq} \mathit{fun}2 \Leftrightarrow \mathcal{Q}\mathit{fun}1 \subseteq \mathcal{Q}\mathit{fun}2 \qquad (2.8)$$

$$\mathit{fun}1 \overset{\mathcal{F}}{\subset} \mathit{fun}2 \Leftrightarrow \mathcal{Q}\mathit{fun}1 \subset \mathcal{Q}\mathit{fun}2 \qquad (2.9)$$

A measure of the *degree of subsetness* defines a weaker form of set inclusion [22,23]. It is only defined for fuzzy sets that are countable i.e. sets of type finite $\mathcal{F}T$.

$$\boxed{\begin{array}{l} [T] \\ \hline \_ \overset{\subseteq}{\sim} \_ : (\text{finite } \mathcal{F}T) \times (\text{finite } \mathcal{F}T) \to \mathcal{M} \\ \hline \forall \mathit{fun}1, \mathit{fun}2 : \text{finite } \mathcal{F}T \ \bullet \ (\mathit{fun}1 \neq \ \text{empty} \ ) \Rightarrow \\ \qquad (\mathit{fun}1 \overset{\subseteq}{\sim} \mathit{fun}2 = \text{count} \ (\mathit{fun}1 \ \text{and} \ \mathit{fun}2)/\text{count} \ (\mathit{fun}1)) \ \wedge \\ \qquad (\mathit{fun}1 = \ \text{empty} \ ) \Rightarrow (\mathit{fun}1 \overset{\subseteq}{\sim} \mathit{fun}2 = 1) \end{array}}$$

2.3.5   Set modifiers and fuzzy numbers

Fuzzy set modifiers transform the membership values of reference elements that exhibit partial membership in a set of type $\mathcal{F}T$. When applied to sets of type $\mathcal{C}T$ they have no effect. They are sometimes referred to as *linguistic hedges* and an analogy has been drawn between them and the action of adjectival modifiers such as *very, generally* etc. in natural language. The meaning ascribed to each and the mathematical description provided have been argued intuitively. The reader is directed to [6,38,39,49] for a detailed discussion of these issues. The definition below is for the *concentration* modifier, very.

$[T]$

very $: (\mathcal{F}T) \rightarrow (\mathcal{F}T)$

$\forall t : T;\ fun : (\mathcal{F}T) \bullet$ very$(fun)(t) = fun(t) * fun(t)$

The toolkit also provides a definition for the *dilation* modifier, somewhat, where the square root of the membership value is used.

Approximation hedges such as *near(m), around(m), roughly(m)* can be modelled as fuzzy numbers. In a fuzzy system where t represents some measured scalar variable, the fuzzy set parameters could be dependent on the magnitude of t alone and scaled accordingly [6]. For example the degree to which *9.5* is *around 10* may be considered to be the same as that to which *95* is *around 100*, *950* is *around 1000* and so on. Differing scaling factors would be used for differing linguistic descriptions.

The following definition models the approximation hedge *near* as a triangular fuzzy set centred about the positive real number m.

---

near $: \mathbb{R}^+ \rightarrow (\mathcal{F}\mathbb{R}^+)$

$\forall p, m : \mathbb{R}^+ \bullet ((p \leq (m - 0.15 * m)) \wedge (p \geq (m + 0.15 * m))$
$\qquad \Rightarrow (\text{near}(m))(p) = 0) \wedge$
$\qquad ((m - 0.15 * m) < p < (m + 0.15 * m)) \Rightarrow$
$\qquad (\text{near}(m))(p) = 1 - abs((m - p)/0.15 * m)$

The toolkit also contains a similar definition for the approximation hedge around. In this case a scaling factor of 0.25 is used to increase the support set of the fuzzy number, suggesting that we would expect more numbers to be around m than near to m. These definitions are provided only as an example and we recognise that the set parameters are subjective and may be dependent on the context in which the hedge is to be used. Gaussian or trapezoid set parameters could be added to the toolkit if necessary. The reader is directed to [2,6] for more detail on possible set parameters and the practical application of approximation hedges of this type in fuzzy systems.

The toolkit provides a generic definition for the fuzzy quantifier *most*. It is based on that in [17].

$$\begin{array}{|l}
\text{most} : \mathcal{M} \to \mathcal{M} \\
\hline
\forall\, m : \mathcal{M} \bullet (0 \leq m < 0.3) \Rightarrow \text{most}(m) = 0 \, \wedge \\
\qquad\qquad (0.3 \leq m < 0.8) \Rightarrow \text{most}(m) = 2 * m - 0.6 \, \wedge \\
\qquad\qquad (m \geq 0.8) \Rightarrow \text{most}(m) = 1
\end{array}$$

2.3.6 Fuzzy set truncation and scaling

The toolkit contains generic definitions for fuzzy set truncation and fuzzy set scaling. The fuzzy set truncation relation $\stackrel{cm}{\Rightarrow}$ restricts the membership of reference set elements to some threshold value. The fuzzy set scaling relation $\stackrel{cp}{\Rightarrow}$ restricts and adjusts set membership to preserve the shape of the fuzzy set. They are sometimes referred to as *correlation minimum* and *correlation product* and are useful as adjustment operators in fuzzy reasoning [6].

$$\begin{array}{|l}
[T] \\
\hline
\_ \stackrel{cm}{\Rightarrow} \_ : (\mathcal{M} \times \mathcal{F}T) \to \mathcal{F}T \\
\hline
\forall\, m : \mathcal{M};\ fun : \mathcal{F}T \bullet \\
\qquad (\forall\, t : \text{dom}\, fun \bullet (m \stackrel{cm}{\Rightarrow} fun)(t) = min\{m, fun(t)\})
\end{array}$$

$$\begin{array}{|l}
[T] \\
\hline
\_ \stackrel{cp}{\Rightarrow} \_ : (\mathcal{M} \times \mathcal{F}T) \to \mathcal{F}T \\
\hline
\forall\, m : \mathcal{M};\ fun : \mathcal{F}T \bullet \\
\qquad (\forall\, t : \text{dom}\, fun \bullet (m \stackrel{cp}{\Rightarrow} fun)(t) = m * fun(t))
\end{array}$$

2.3.7 Fuzzy relations

The generic symbols, $\stackrel{\mathcal{F}}{\leftrightarrow}$ and $\stackrel{\mathcal{C}}{\leftrightarrow}$, define fuzzy and crisp relations.

$$X \stackrel{\mathcal{F}}{\leftrightarrow} Y == \mathcal{F}(X \times Y)$$
$$X \stackrel{\mathcal{C}}{\leftrightarrow} Y == \mathcal{C}(X \times Y)$$

A *identity relation* can be defined, using the extended notation.

$$\begin{array}{|l}
[X] \\
\hline
\stackrel{\mathcal{F}}{\text{id}} : \mathcal{F}X \to (X \stackrel{\mathcal{F}}{\leftrightarrow} X) \\
\hline
\forall\, fun : \mathcal{F}X;\ x_1, x_2 : X \bullet \\
\qquad (x_1 = x_2) \Rightarrow (\stackrel{\mathcal{F}}{\text{id}}\, fun)(x_1 \mapsto x_2) = (\text{total}\, fun)(x_1) \, \wedge \\
\qquad (x_1 \neq x_2) \Rightarrow (\stackrel{\mathcal{F}}{\text{id}}\, fun)(x_1 \mapsto x_2) = 0
\end{array}$$

This has the same meaning using either the extended or conventional notation.

$$Q(\overset{\mathcal{F}}{\text{id}} \; xset) = \text{id}(Q(xset)) \tag{2.10}$$

where $xset$ is of type CX. This definition for the identity relation is based on that found in [26].

A *fuzzy relational inverse* can also be defined.

$$\boxed{\begin{array}{l} =[X, Y] \\ \hline \quad \_^{-\mathcal{F}} : (X \overset{\mathcal{F}}{\leftrightarrow} Y) \to (Y \overset{\mathcal{F}}{\leftrightarrow} X) \\ \hline \quad \forall\, R : (X \overset{\mathcal{F}}{\leftrightarrow} Y);\; x : X;\; y : Y \bullet R^{-\mathcal{F}}(y \mapsto x) = R(x \mapsto y) \end{array}}$$

This has the same meaning using either the extended or conventional notation. $\forall\, R : X \overset{\mathcal{C}}{\leftrightarrow} Y$,

$$Q(R^{-\mathcal{F}}) = (QR)^{-1} \tag{2.11}$$

A *fuzzy cartesian product* can be formed between any two fuzzy sets [7]. Membership of any maplet $x \mapsto y$ in the resulting fuzzy relation is the minimum of the membership of x and y in the original fuzzy sets.

$$\boxed{\begin{array}{l} =[X, Y] \\ \hline \quad \_\overset{\mathcal{F}}{\times}\_ : (\mathcal{F}X \times \mathcal{F}Y) \to (X \overset{\mathcal{F}}{\leftrightarrow} Y) \\ \hline \quad \forall\, x : X;\; y : Y;\; xset : \mathcal{F}X;\; yset : \mathcal{F}Y \bullet \\ \qquad (xset \overset{\mathcal{F}}{\times} yset)(x \mapsto y) = min\{(\text{total } xset)(x), (\text{total } yset)(y)\} \end{array}}$$

When applied to sets of type CX and CY it has the same meaning as the cartesian product $\times$. $\forall\, xset : CX;\; yset : CY$,

$$Q(xset \overset{\mathcal{F}}{\times} yset) = Q(xset) \times Q(yset) \tag{2.12}$$

2.3.8 Range and Domain for a fuzzy relation

The domain(dom) and range(ran) for fuzzy relations can be defined as a fuzzy set [7]. The membership of an element x in the domain (and an element y in the range) could be considered to be equal to the maximum of all memberships of the mappings $\{x\} \to Y$ (or $X \to \{y\}$) in the fuzzy relation, $X \overset{\mathcal{F}}{\leftrightarrow} Y$. The definition for dom is shown below.

$$
\begin{array}{|l}
\hline
\!=[X, Y] \!= \\
\quad \overset{\mathcal{F}}{\mathrm{dom}} \colon (X \overset{\mathcal{F}}{\leftrightarrow} Y) \to \mathcal{F}X \\
\hline
\quad \forall R : (X \overset{\mathcal{F}}{\leftrightarrow} Y);\ x : X \bullet (\overset{\mathcal{F}}{\mathrm{dom}}\ R)(x) = max\{y : Y \bullet (\mathsf{total}\ R)(x \mapsto y)\} \\
\hline
\end{array}
$$

This has the same meaning for the domain of a crisp relations written either in the conventional or extended notation. $\forall R : X \overset{\mathcal{C}}{\leftrightarrow} Y$,

$$
\mathcal{Q}(\overset{\mathcal{F}}{\mathrm{dom}}\ R) = \mathrm{dom}(\mathcal{Q}(R)) \tag{2.13}
$$

2.3.9 Range and domain restrictions (and anti- restriction) for fuzzy relations

A *domain* and *range restriction* for a fuzzy relation can also be defined. The membership of the maplet $x \mapsto y$ in the restricted fuzzy relation is given by the minimum of the original membership of $x \mapsto y$ in the relation, R and the membership of x or y in the restricting set, *xset* or *yset*.

$$
\begin{array}{|l}
\hline
\!=[X, Y]\!= \\
\quad \_\overset{\mathcal{F}}{\vartriangleleft}\_ : (\mathcal{F}X) \times (X \overset{\mathcal{F}}{\leftrightarrow} Y) \to (X \overset{\mathcal{F}}{\leftrightarrow} Y) \\
\quad \_\overset{\mathcal{F}}{\vartriangleright}\_ : (X \overset{\mathcal{F}}{\leftrightarrow} Y) \times (\mathcal{F}Y) \to (X \overset{\mathcal{F}}{\leftrightarrow} Y) \\
\hline
\quad \forall x : X;\ y : Y;\ R : (X \overset{\mathcal{F}}{\leftrightarrow} Y);\ xset, yset : (\mathcal{F}X) \bullet \\
\quad (xset \overset{\mathcal{F}}{\vartriangleleft} R)(x \mapsto y) = min\{(\mathsf{total}\ R)(x \mapsto y), (\mathsf{total}\ xset)(x)\} \wedge \\
\quad (R \overset{\mathcal{F}}{\vartriangleright} yset)(x \mapsto y) = min\{(\mathsf{total}\ R)(x \mapsto y), (\mathsf{total}\ yset)(y)\} \\
\hline
\end{array}
$$

These have the same meaning for crisp relations written either in the conventional or extended notation. $\forall R : X \overset{\mathcal{C}}{\leftrightarrow} Y$, $xset : \mathcal{C}X$, $yset : \mathcal{C}Y$,

$$
\mathcal{Q}(xset \overset{\mathcal{F}}{\vartriangleleft} R) = \mathcal{Q}(xset) \vartriangleleft \mathcal{Q}(R) \tag{2.14}
$$

$$
\mathcal{Q}(R \overset{\mathcal{F}}{\vartriangleright} yset) = \mathcal{Q}(R) \vartriangleright \mathcal{Q}(yset) \tag{2.15}
$$

The toolkit also defines a *domain* and *range anti-restriction* for a fuzzy relation. The membership of the maplet $x \mapsto y$ in the anti-restricted fuzzy relation is given by the minimum of the original membership of $x \mapsto y$ in the relation, $\mathsf{total}\ R$ and the membership of x or y in the *complement* of the anti-restricting set, *xset* or *yset*. The definitions are very similar to those for $\overset{\mathcal{F}}{\vartriangleleft}$ and $\overset{\mathcal{F}}{\vartriangleright}$, with $(\mathsf{not}\ (\mathsf{total}\ xset))(x)$ and $(\mathsf{not}\ (\mathsf{total}\ yset))(y)$ replacing $(\mathsf{total}\ xset)(x)$ and $(\mathsf{total}\ yset)(y)$ respectively.

2.3.10 The *max-min* relational composition operator for fuzzy relations

$$
\begin{array}{l}
\underline{\quad}[X, Y, Z]\underline{\quad\quad\quad\quad\quad\quad\quad\quad\quad\quad\quad\quad\quad\quad} \\[4pt]
\quad \underline{\ \ }\overset{\mathcal{F}}{\,\fatsemi\,}\underline{\ \ } : (X \overset{\mathcal{F}}{\leftrightarrow} Y) \times (Y \overset{\mathcal{F}}{\leftrightarrow} Z) \to (X \overset{\mathcal{F}}{\leftrightarrow} Z) \\[4pt]
\quad \underline{\ \ }\overset{\mathcal{F}}{\circ}\underline{\ \ } : (Y \overset{\mathcal{F}}{\leftrightarrow} Z) \times (X \overset{\mathcal{F}}{\leftrightarrow} Y) \to (X \overset{\mathcal{F}}{\leftrightarrow} Z) \\[4pt]
\underline{\quad\quad\quad\quad\quad\quad\quad\quad\quad\quad\quad\quad\quad\quad\quad\quad\quad} \\[4pt]
\forall R_1 : (X \overset{\mathcal{F}}{\leftrightarrow} Y);\ R_2 : (Y \overset{\mathcal{F}}{\leftrightarrow} Z);\ x : X;\ z : Z \bullet (R_1 \overset{\mathcal{F}}{\,\fatsemi\,} R_2)(x \mapsto z) \\[4pt]
\quad = max\{y : Y \bullet min\{(\text{total } R_1)(x \mapsto y), (\text{total } R_2)(y \mapsto z)\}\}\ \wedge \\[4pt]
\quad (R_1 \overset{\mathcal{F}}{\,\fatsemi\,} R_2)(x \mapsto z) = (R_2 \overset{\mathcal{F}}{\circ} R_1)(x \mapsto z)
\end{array}
$$

There are two common composition operators for fuzzy relations, *max-min* and *min-max* [2]. Only *max-min* is shown here. The *min-max* operator would be defined by interchanging *min* and *max* in the above definition.

This definition has the same meaning for crisp relations written in either the conventional or extended notation. $\forall R_1 : X \overset{\mathcal{C}}{\leftrightarrow} Y$ and $R_2 : Y \overset{\mathcal{C}}{\leftrightarrow} Z$,

$$\mathcal{Q}(R_1 \overset{\mathcal{F}}{\,\fatsemi\,} R_2) = \mathcal{Q}(R_1) \,\fatsemi\, \mathcal{Q}(R_2) \tag{2.16}$$

$$\mathcal{Q}(R_2 \overset{\mathcal{F}}{\circ} R_1) = \mathcal{Q}(R_2) \circ \mathcal{Q}(R_1) \tag{2.17}$$

2.3.11 A *fuzzy relational image* for fuzzy relations

$$
\begin{array}{l}
\underline{\quad}[X, Y]\underline{\quad\quad\quad\quad\quad\quad\quad\quad\quad\quad\quad\quad\quad\quad\quad} \\[4pt]
\quad \underline{\ \ }\overset{\mathcal{F}}{(\!|}\ \underline{\ \ }\ \overset{\mathcal{F}}{|\!)} : (X \overset{\mathcal{F}}{\leftrightarrow} Y) \times \mathcal{F}X \to \mathcal{F}Y \\[4pt]
\underline{\quad\quad\quad\quad\quad\quad\quad\quad\quad\quad\quad\quad\quad\quad\quad\quad\quad} \\[4pt]
\forall R : (X \overset{\mathcal{F}}{\leftrightarrow} Y);\ set : \mathcal{F}X;\ y : Y \bullet \\[4pt]
\quad (R \overset{\mathcal{F}}{(\!|}\ set\ \overset{\mathcal{F}}{|\!)})(y) = max\{x : X \bullet min\{(\text{total } R)(x \mapsto y), (\text{total } set)(x)\}\}
\end{array}
$$

This definition is based on that given in [26] and has the same meaning for crisp relations and sets using either the conventional or extended notation. $\forall R : X \overset{\mathcal{C}}{\leftrightarrow} Y, set : \mathcal{C}X$ and $y : Y$,

$$\mathcal{Q}(R \overset{\mathcal{F}}{(\!|}\ set\ \overset{\mathcal{F}}{|\!)}) = \mathcal{Q}(R)(\!|\ \mathcal{Q}(set)\ |\!) \tag{2.18}$$

2.3.12 Fuzzy functions

A series of fuzzy functions are defined in the toolkit. Each is a generalisation of the corresponding definition in the conventional notation [41]. Crisp versions of these are also be defined and it can be shown that they have the same meaning in either the conventional or extended notation [29].

A *fuzzy partial function.*

$$X \xrightarrow{\mathcal{F}} Y == \{R : X \xleftrightarrow{\mathcal{F}} Y \mid \forall x : X; \ y_1, y_2 : Y \bullet \\ (R(x \mapsto y_1) > 0) \wedge (R(x \mapsto y_2) > 0) \Rightarrow y_1 = y_2\}$$

A *fuzzy total function*

$$X \xrightarrow{\mathcal{F}} Y == \{R : X \xrightarrow{\mathcal{F}} Y \mid \forall x : X \bullet (\mathrm{dom}\ R)(x) > 0\}$$

A *fuzzy partial injection*

$$X \xrightarrowtail{\mathcal{F}} Y == \{R : X \xrightarrow{\mathcal{F}} Y \mid \forall x_1, x_2 : X; \ y : Y \bullet \\ (R(x_1 \mapsto y) > 0 \wedge R(x_2 \mapsto y) > 0) \Rightarrow x_1 = x_2\}$$

A *fuzzy total injection*

$$X \xrightarrowtail{\mathcal{F}} Y == \{R : X \xrightarrow{\mathcal{F}} Y \mid \forall x_1, x_2 : X; \ y : Y \bullet \\ (R(x_1 \mapsto y) > 0 \wedge R(x_2 \mapsto y) > 0) \Rightarrow x_1 = x_2\}$$

A *fuzzy partial surjection*

$$X \xrightarrow{\mathcal{F}} \hspace{-1.2em}\rightarrow Y == \{R : X \xrightarrow{\mathcal{F}} Y \mid \forall y : Y \bullet (\mathrm{ran}\ R)(y) > 0\}$$

A *fuzzy total surjection*

$$X \xrightarrow{\mathcal{F}} \hspace{-1.2em}\rightarrow Y == \{R : X \xrightarrow{\mathcal{F}} Y \mid \forall y : Y \bullet (\mathrm{ran}\ R)(y) > 0\}$$

A *fuzzy bijection*

$$X \xrightarrowtail{\mathcal{F}} Y == \{R : X \xrightarrowtail{\mathcal{F}} Y \mid \forall y : Y \bullet (\mathrm{ran}\ R)(y) > 0\}$$

2.4   Some Illustrative Examples

This section explores possible ways in which the toolkit can be used as a modelling tool for the problem types of interest. The intention is not to present a complete model, but instead to use parts of each example to illustrate the usefulness of the representation scheme.

2.4.1 Invariant fuzzy sets

Concepts that we are unable to precisely categorise into distinct classes or types can be described linguistically and represented using fuzzy sets. The toolkit allows us to model such concepts in a Z specification. Those fuzzy sets whose parameters remain unchanged throughout a specification are referred to as being *invariant*. The definitions for such sets can appear as a series of axiomatic definitions at the beginning of a specification. As a simple illustration consider the concept of *profit levels* in an organisation. Profits levels could be interpreted as a linguistic variable whose values can be described using fuzzy sets. Using the fuzzy logic toolkit we could model this as follows.

$$profits : \mathbb{P}\,\mathbb{R}^+$$

$$highProfits : \mathcal{F}\,profits$$
$$averageProfits : \mathcal{F}\,profits$$
$$lowProfits : \mathcal{F}\,profits$$

The mapping functions which deliver set membership would be defined as constraints in the axiomatic definitions. For example, *averageProfits* could be defined using the fuzzy number, around.

$$averageProfits : \mathcal{F}\,profits$$
$$averageProfits = \mathsf{around}(100000)$$

Similar constraints could developed for *lowProfits* and *highProfits*. There are many other examples of linguistic variables that could be modelled as invariant fuzzy sets. These include concepts such as *customer demand* in a marketing system, *student enrolments* in an educational system, *treatment costs* in a health system and so on. In each case the reference set is numeric and the mapping functions that deliver set membership will need to be determined.

2.4.2 Multidimensional fuzzy sets

A *multidimensional* fuzzy set is one where set membership may be due to more than one property of the reference set [11]. Unlike the sets described above there is no mapping function to deliver set membership, but instead the reference set element will be fully or partially added or removed as a result of a system operation. To illustrate this we will extend the example and consider the employees within an organisation. As well as being interested in whether a person is employed by the organisation we are also interested in the motivation, productivity and experience of the workforce. These can be seen as graded concepts and thus be modelled as fuzzy sets. If we assume the basic type, *person*, in our model then we could write a state schema as follows.

[*person*]

```
_ staffing _____
  employed : C person
  motivated : F person
  experienced : F person
  productive : F person
```

For simplicity employment is being modelled as a crisp concept i.e. a person is either employed or they are not employed. If there was the possibility of partial employment or if we were uncertain as to the interpretation of membership in this set, then it could be modelled as a fuzzy set. It is also assumed that in the initial system state all of these sets are empty. In other words the degree of membership of all reference set elements in each of these sets is zero.

```
_ INIT _____
  staffing
  _____
  employed = empty
  motivated = empty
  experienced = empty
  productive = empty
```

The set operators and modifiers defined in the fuzzy logic toolkit are useful in modelling approximate statements or heuristics that describe aspects of the problem domain. For example

- *highly motivated employees* could be modelled as the set,
  very(*motivated* and *employed*)
- *inexperienced employees* could be modelled as the set,
  not (*experienced*) and *employed*
- *experienced and motivated people are the most prodcutive* could be modelled as *productive = experienced* and *motivated*

Some of these could be built into the specification. For example we could incorporate the heuristic that *the most productive people are those that are motivated and experienced*, into the Δ schema as follows.

```
_ Δstaffing _____
  staffing
  staffing'
  _____
  productive' = productive ⊕ (motivated' and experienced)
```

System operations can be written to adjust set membership in each of these fuzzy sets. Adjustment to zero is equivalent to element removal and adjustment to one is equivalent to element addition. The schema below adjusts the motivation for a group of employees.

```
__ newMotivation _____
 Δstaffing
 employeeGroup? : person ⇸ M
 _____
 ∀ p : dom employeeGroup? • p in employed
 motivated' = motivated ⊕ employeeGroup?
 employed' = employed
 experienced' = experienced
```

It can be shown that empty is a *finite fuzzy* set. Thus in the initial system state *employed, motivated, experienced* and *productive* are all finite fuzzy sets. Provided the overriding function is finite then the sets *motivated, employed, experienced* and *productive* will remain finite fuzzy sets throughout subsequent system states. This is important because it means that they are countable and properties such as the degree of fuzziness, set similarity and degree of subsetness can be evaluated. As will be shortly such properties give us the ability to estimate the degree to which a set of requirements might be met.

2.4.3 A soft pre-condition

This section explores the possibility that a schema pre-condition may only partially hold and is motivated by the work on soft functional requirements [25]. A pre-condition is a constraint placed on an operation. The operation can only succeed if the pre-condition holds. We have already introduced the concepts of employment and profit levels. The following operational schema adds a person to the set of employed people only when profit levels are high.

```
__ employ _____
 Δstaffing
 p? : person
 profitLevel? : profits
 _____
 profitLevel? in highProfits
 p? notin employed
 employed' = employed ⊕ {p? ↦ 1}
 motivated' = motivated
 experienced' = experienced
```

Profit levels and employment are modelled in crisp terms, and the operation only succeeds when *profitLevel?* exhibits full membership in the fuzzy set *highProfits* and the membership of *p?* in the set *employed* is zero.

However profit levels could be considered to be a fuzzy concept and thus it may be possible to determine the *degree* to which profit levels are high. Consequently the pre-condition may only *partially* hold and it could be considered to be a 'soft' pre-condition.

There are several ways to model this. For example we could consider that the degree to which we employ a person is related to the degree to which profits are high. The schema below takes this approach.

---
employ1
$\Delta staffing$
$p? : person$
$profitLevel? : profits$

$p?$ notin $employed$
$employed' = employed \oplus$
$\{p? \mapsto highProfits(profitLevel?)\}$
$motivated' = motivated$
$experienced' = experienced$

The pre-condition that the person is not already employed is retained and the set *employed* is now fuzzy i.e. it has been declared to be of type $\mathcal{F}person$ rather than $\mathcal{C}person$ as above. In this approach the degree to which the pre-condition holds directly determines the degree of membership of the element in the fuzzy set.

An alternative approach is to suggest that the degree to which the pre-condition holds determines the degree to which the operation succeeds. To illustrate this imagine a group of people whose employment status is affected by the degree to which the profit levels are high. This group may include people already employed and/or people seeking employment. The schema below attempts to model an adjustment in the employment mix of the people concerned as a result of the profit levels within an organisation. The effect on the employment status of a particular individual is not specified, just the overall effect on the group.

---
employ2
$\Delta staffing$
$P? : person \nrightarrow \mathcal{M}$
$profitLevel? : profits$

$\text{dom } P? \lhd employed' = \text{dom } P? \lhd employed$
$((\text{dom } P? \lhd employed') \approx P?) = highProfits(profitLevel?)$
$motivated' = motivated$
$experienced' = experienced$

The behaviour of this schema is discussed in some detail in [28]. However some general comments can be made here.

It is assumed that the set *employed* is of type $\mathcal{F}$*person* and has been initialised to the empty fuzzy set, empty . Thus *employed* is defined for all elements of the reference set *person* and the domain of *P*? is a subset of the domain of *employed*. The domain anti-restriction operation on the set, *employed* ensures that the membership of those reference set elements not in *P*? are unchanged as a result of the operation. It can be shown that the set dom *P*? ◁ *employed'* is a finite fuzzy set and thus the degree of equality measure can be used between it and *P*?. It can also be shown that the operation succeeds totally when *highProfits*(*profitLevel*?) = 1 i.e. dom *P*? ◁ *employed* = *P*?. Finally, when the partial function *P*? contains only one reference set element it can be shown that operation behaves in the same way as that described by the schema *employ1* [28,29].

2.4.4   Fuzzy and crisp relations

To illustrate the use of fuzzy and crisp relations we will introduce a second and more detailed example. The example models part of a health care network. There are a series of institutions that deliver a variety of medical services to the general public. The model that follows relates to service delivery only.There are five basic types of interest.

[*institution*, *client*, *medicalServiceType*, *date*]

A *medicalService* is the use of a *medicalServiceType* on a particular date.

medicalService == *medicalServiceType* × *date*

ClientService is a state schema describing the service delivery of the system.

$$
\begin{array}{l}
\rule{0pt}{1em}\textit{ClientService} \\
\hline
\textit{offeredBy} : \textit{medicalService} \stackrel{\mathcal{C}}{\leftrightarrow} \textit{institution} \\
\textit{highQuality} : \textit{medicalService} \stackrel{\mathcal{F}}{\leftrightarrow} \textit{institution} \\
\textit{usedBy} : \textit{client} \stackrel{\mathcal{C}}{\leftrightarrow} \\
\qquad (\textit{medicalService} \stackrel{\mathcal{C}}{\leftrightarrow} \textit{institution}) \\
\textit{satisfaction} : \textit{client} \stackrel{\mathcal{F}}{\leftrightarrow} \\
\qquad (\textit{medicalService} \stackrel{\mathcal{C}}{\leftrightarrow} \textit{institution}) \\
\hline
\textit{highQuality} \stackrel{\mathcal{F}}{\subseteq} \textit{offeredBy} \\
\textit{satisfaction} \stackrel{\mathcal{F}}{\subseteq} \textit{usedBy}
\end{array}
$$

The fact that a medical service is offered by an institution has been modelled as a crisp relation. There is no fuzziness here — either the medical service is offered or it is not. If we were modelling whether a medical service *type* was offered by an institution then perhaps some fuzziness would be permitted. This might allow us to represent the idea of an institution offering a particular type of medical service *most* of the time, or *some* of the time and so on.

We are interested in the *quality* of medical services offered by the various institutions throughout the network. Degrees of service quality are possible and consequently it has been modelled as a fuzzy relation. For simplicity we have only included a single relation, namely *highQuality*, but if concepts such as *averageQuality* and *lowQuality* were important then they could be modelled as well. A similar approach has been taken when dealing with the use of medical services offered throughout the system and client satisfaction with those service offerings. The schema constraints state that the degree of membership of a *medicalService* offered by a particular *institution* in the relation *highQuality* has to be less than or equal to that in the *offeredBy*. In other words we are only interested in the quality of those services offered by institutions in the health care network. Similarly we are only interested in the satisfaction of clients that use the medical services offered by the institutions in the network.

An advantage of modelling concepts such as service quality and client satisfaction in fuzzy terms is that we are able to ask softer questions of the specification. By a 'soft' question we mean one to which there is no definitive answer i.e where it is difficult to simply answer *yes* or *no*. Instead the answer could be expressed in approximate or more general terms. For example, rather than asking whether the medical services offered throughout the network are of high quality, we could ask *to what degree* are all medical services offered by the institutions of a high quality? More generally we could ask to what degree are *most* medical services of a high quality and so on. In the next section we will show how questions of this type can be answered and how they could be re-stated as a series of quantified propositions that express part of the system requirements.

2.4.5   Linguistically quantified propositions

When *highQuality* = *offeredBy* all offered services are of the highest quality i.e.

$$\forall ms : medicalService; \ i : institution \bullet$$
$$offeredBy(ms \mapsto i) = 1 \Leftrightarrow$$
$$highQuality(ms \mapsto i) = 1 \ \wedge$$
$$\quad offeredBy(ms \mapsto i) = 0 \Leftrightarrow$$
$$highQuality(ms \mapsto i) = 0$$

The *degree* to which *all* offered services are of a high quality could be given by the degree to which the two relations are equal i.e.

$highQuality \approx offeredBy$

The following schema determines the quality of *all* offered services across the system

---
__ *qualityOfOfferings*1 _____
Ξ *ClientService*
$m! : \mathcal{M}$

$offeredBy \neq$ empty
$m! = (highQuality \approx offeredBy)$

The following schema determines the quality of *most* offered services across the system

---
__ *qualityOfOfferings*2 _____
Ξ *ClientService*
$m! : \mathcal{M}$

$offeredBy \neq$ empty
$m! = \mathsf{most}(highQuality \approx offeredBy)$

The fuzzy quantifier most is defined in section 2.3. The definition always ensures that the degree to which *most* offered services are of a high quality is always greater or equal to that of *all* offered services. The schema precondition ensures that at least one service has to be offered somewhere in the system before the question(s) can be answered.

Using appropriate fuzzy quantifiers and linguistic hedges a series of similar questions could be framed. For example

- To what degree are *most* clients satisfied with the medical services they have used within the system?
  i.e. $\mathsf{most}(satisfaction \approx usedBy)$
- To what degree are the services offered across the system of *very* high quality?
  i.e. $(\mathsf{very}\ highQuality) \approx offeredBy$
- To what degree are most clients somewhat satisfied with the medical services they have used within the system?
  i.e. $\mathsf{most}((\mathsf{somewhat}\ satisfaction) \approx usedBy)$

Questions of this type could be re-stated as series of *linguistically quantified propositions* [17]. For example

> Most clients are satisfied
> All services are of a very high quality

Most clients are somewhat satisfied
Most services are of a high quality
etc.

All of these are fuzzy in the sense that a degree of truth for each proposition can be evaluated. They could be interpreted as a statement of requirements for the system and provide a way of determining to what extent the system meets these requirements at any given time.

2.4.6 Modelling conflict and agreement

In a recent paper Z. Pawlak [34] observed that

It seems that fuzzy and rough sets are perfect candidates for modelling conflict situations in the presence of uncertainty, but to my knowledge not very much has been done in this area so far, particularly when fuzzy sets are concerned.

Pawlak presents a framework for conflict analysis based on rough set theory[5]. However concepts such as *support*, *agreement* and *conflict* which form part of his analysis could also be treated in fuzzy terms. Given that some of the system problems of interest could include conflict and contradiction it may be possible to build such aspects of this into a formal model. To illustrate how this might be achieved a third example is introduced.

In this case the example is somewhat abstract. Imagine a set of *agents* (consisting perhaps of people, institutions, organisational units, stakeholders etc) and a set of *issues* that are important to these agents. We could define the following fuzzy relations

$$support : agents \overset{\mathcal{F}}{\leftrightarrow} issues$$
$$inAgreement : (agents \times agents) \overset{\mathcal{F}}{\leftrightarrow} issues$$

The relation, *support*, indicates the degree of support that an agent expresses for a particular issue. The relation, *inAgreement* indicates the extent to which any two agents are in agreement over a particular issue. Obviously any agent has to be in total agreement with themselves on all issues.

For any agent a, the fuzzy relational image $support$ $(\mathcal{P}\{a\})$ is a fuzzy set of type $\mathcal{F}issues$ indicating the degree of support of that particular agent a for each issue.

We will now define *inAgreement* in terms of *support*. It is assumed that the degree to which any two agents are in agreement on a particular issue is

---

[5] A *rough set* is a crisp set that is not precisely defined. There is no concept of partial set membership but instead the idea that an element may *surely* or *possibly* belong to the set. Roughness arises from a lack of information rather than any natural imprecision in the problem domain.

related to the 'closeness' of their individual support for that issue. A simple closeness measure for any $m_1, m_2 : \mathcal{M}$ could be the absolute value of the difference between m_1 and m_2 i.e. $abs(m_1 - m_2)$.

The relation, $inAgreement$ can be defined as follows

$$\forall a_1, a_2 : agents; \; i : issues \bullet$$
$$inAgreement((a_1, a_2) \mapsto i) =$$
$$1 - abs(support(a_1 \mapsto i) - support(a_2 \mapsto i))$$

The definition has the following properties.

1. $inAgreement((a_1, a_1) \mapsto i) = 1$
2. $inAgreement((a_1, a_2) \mapsto i) = inAgreement((a_2, a_1) \mapsto i)$
3. $inAgreement((a_1, a_2) \mapsto i) = 1 \wedge inAgreement((a_2, a_3) \mapsto i) = 1 \Rightarrow inAgreement((a_1, a_3) \mapsto i) = 1$

A further relation, $inConflict$ can be defined in terms of $inAgreement$ i.e.

$$inConflict : (agents \times agents) \overset{\mathcal{F}}{\leftrightarrow} issues \bullet inConflict = \mathsf{not} \; inAgreement$$

Note that the concepts of *agreement* and *conflict* could have been modelled the other way around i.e. $inConflict((a_1, a_2) \mapsto i) = abs(support(a_1 \mapsto i) - support(a_2 \mapsto i))$ and $inAgreement = \mathsf{not} \; inConflict$. The modelling of these as fuzzy concepts permits the possibility of overlap. In other words there may be some degree of agreement and conflict on certain issues at the same time.

The following describes a system state which formalises these ideas.

$$[agents, issues]$$

$conflictModel$

$support : agents \overset{\mathcal{F}}{\leftrightarrow} issues$
$inAgreement : (agents \times agents) \overset{\mathcal{F}}{\leftrightarrow} issues$
$inConflict : (agents \times agents) \overset{\mathcal{F}}{\leftrightarrow} issues$

$\forall a_1, a_2 : agents; \; i : issues \bullet$
$\quad inAgreement((a_1, a_2) \mapsto i) =$
$\quad\quad 1 - abs(support(a_1 \mapsto i) - support(a_2 \mapsto i))$
$inConflict = \mathsf{not} \; inAgreement$

If the sets of *agents* and *issues* were non-empty and finite (i.e. they were known before the specification was developed) then they could be declared as free types i.e.

$$agents ::= a_1 \mid a_2 \mid ...a_{n-1} \mid a_n$$
$$issues ::= i_1 \mid i_2 \mid ...i_{m-1} \mid i_m$$

We might choose to define the initial state for the system in several ways, depending on how we view the problem. For example we might consider that initially there is no support by any of the agents for the issues of interest. In this case the schema describing the initial state would be as follows.

```
┌─ INIT ────────────────────────────────────
│  conflictModel
│ ──────────────────────────────
│  support = empty
└────────────────────────────────────────────
```

where $inAgreement = \mathcal{U}$ and $inConflict = $ empty

Alternatively we might consider that there is full support for all issues. The $INIT$ schema would then be written as

```
┌─ INIT ────────────────────────────────────
│  conflictModel
│ ──────────────────────────────
│  support = \mathcal{U}
└────────────────────────────────────────────
```

where $inAgreement = $ empty and $inConflict = \mathcal{U}$

The specification would provide an operation to allow an *agent* to express some degree of support for a particular issue. This will be a function override on the fuzzy relation, *support* i.e.

```
┌─ expressSupport ──────────────────────────
│  ΔconflictModel
│  a? : agents
│  i? : issues
│  supportLevel : M
│ ──────────────────────────────
│  support' = support ⊕ {(a? ↦ i?) ↦ supportLevel}
└────────────────────────────────────────────
```

We could ask for the degree of *broad agreement* amongst any two agents across *all* issues of interest.

---
broadAgreement
Ξ *conflictModel*
agent1?, agent2? : agents
$m! : \mathcal{M}$

$\overset{\mathcal{F}}{\text{dom}}\ support(agent1?) > 0 \ \wedge \ \overset{\mathcal{F}}{\text{dom}}\ support(agent2?) > 0$

$m! = \overset{\mathcal{F}}{support}\ (\!|\ \overset{\mathcal{F}}{\mathcal{P}}\{agent1\}\ |\!)\ \approx\ \overset{\mathcal{F}}{support}\ (\!|\ \overset{\mathcal{F}}{\mathcal{P}}\{agent2\}\ |\!)$

This might be interpreted as follows. The degree to which any two agents are in broad agreement across a range of issues of interest is equal to the degree of equality (or similarity) between the two fuzzy sets, of type *issues*, that indicate the degree of support each agent shows for all issues of interest. The pre-condition ensures that each agent shows some support for at least one issue. This enquiry can only be made if in the initial state, *support* = empty . This ensures that *support* is finite and that the fuzzy relational images are finite and therefore countable.

The degree of fuzziness of the relation *support* (i.e fuzzyEntropy(*support*)) could be interpreted as indicating the lack of certainty that the *agents* have in expressing their support (or non-support) for the issues of interest. As the degree of fuzziness approaches zero then we might say that there is a more clear cut indication of support or non-support. A degree of fuzziness measure, fuzzyEntropy is defined in section 2.3.

Finally it should be noted that there are many other problem domains where the approaches discussed in this section might be useful. For example a similar approach could be used to model the extent to which a set of system requirements are met by a series of competing vendors or the extent to which a set of criteria are met by a series of tendering organisations and so on. The application of a fuzzy approach to problems involving group and multi-criteria decision making is well established [17,46,18].

2.5   Toolkit Usage — Reflections and Guidelines

A major argument in section 2.1 is that are there some system problems that are not naturally understood in precise or exact terms. Instead they may be characterised by imprecision and uncertainty, and any models that we build to represent them need to take this into account. Fuzziness or naturally occuring imprecision is one major source of uncertainty in such systems, particularly those that are designed to support human decision making or judgement. If conventional Z is being used in the specification of such systems then the specifier would need to define their own strategy for representing inherently fuzzy concepts. The fuzzy logic toolkit for Z is designed to provide a 'standard' approach by providing a mechanism for representing and manipulating fuzzy sets.

Section 2.4 presented a series of examples intended to illustrate the usefulness of the toolkit when modelling the system problems of interest. This section will now reflect on some of the more general issues that arose during the formulation of these examples. In particular it will focus on the choices facing a specifier when

- writing operational schema for a fuzzy set.
- declaring crisp and fuzzy concepts within the same specification.
- defining the initial state of a fuzzy set.

2.5.1   Operational Schema and set declaration

The philosophy behind the use of the toolkit is that the structure $\mathcal{F}basicType$ is a fundamental building block of a specification. The toolkit defines the concepts of emptiness and finiteness through the definitions of an empty fuzzy set (empty) and a finite fuzzy set (finite $\mathcal{F}$). Sets that are declared of type $\mathcal{F}basicType$ are given some initial state which is typically that they are 'empty'. Adding or removing reference set elements is acheived by simply adjusting set membership. Adjustment to one is interpreted as 'adding' a reference set element to the set and adjustment to zero is interpreted as 'removing' that element. Adjustment to values between zero and one can be interpreted as either partial removal or partial addition of an element, depending the direction of the adjustment. Consequently the specification requires only one operational schema, rather than two, for set membership manipulation. Furthermore, unlike in conventional Z, there is no requirement for a pre-condition which checks for set membership or set non-membership prior to removing or adding an element to a set. In this sense a specification written using the toolkit has less constraints than that written in conventional Z. For example

Using the fuzzy logic toolkit

Using conventional Z

$[basic\,Type]$

```
┌─ state ──────────
│ fset : F basic Type
```

```
┌─ INIT ──────────
│ state
│
│ fset = empty
```

```
┌─ setAdjust ──────────
│ Δstate
│ b? : basic Type
│ m? : M
│
│ fset' = fset ⊕ {b? ↦ m?}
```

$[basic\,Type]$

```
┌─ state ──────────
│ set : ℙ basic Type
```

```
┌─ INIT ──────────
│ state
│
│ set = ∅
```

```
┌─ setAdd ──────────
│ Δstate
│ b? : basic Type
│
│ b? ∉ set
│ set' = set ∪ {b?}
```

```
┌─ setRemove ──────────
│ Δstate
│ b? : basic Type
│
│ b? ∈ set
│ set' = set \ {b?}
```

In the above example the override operator, $\oplus$, ensures that any existing membership value for the input variable $b?$ is replaced by the input membership value $m?$. It is functioning as both a 'set union' and 'set difference' operator, depending on the relative values of the existing membership value and of the desired membership value.

It should be noted that if the fuzzy set union operator (or) was used then $fset' = fset$ or $\{b? \mapsto m?\}$ can never result in a decrease in the membership of $b?$ in $fset$, and if the fuzzy set difference operator $(\overset{\mathcal{F}}{\backslash})$ was used then $fset' = fset \overset{\mathcal{F}}{\backslash} \{b? \mapsto m?\}$ can never result in an increase in the membership of $b?$ in $fset$. The specifier therefore has a choice. If he/she wishes to guarantee no decrease in membership as a result of the operation then $setAdjust$ could become

```
┌─ setAdjust ──────────────────────────────────
│ Δstate
│ b? : basic Type
│ m? : M
│
│ fset' = fset or {b? ↦ m?}
```

Conversely if they wish to guarantee no increase in membership as a result of the operation then $setAdjust$ could become

$$
\boxed{\begin{array}{l}
\underline{\;setAdjust\;} \\
\Delta state \\
b? : basicType \\
m? : \mathcal{M} \\
\hline
fset' = fset \overset{\mathcal{F}}{\setminus} \{b? \mapsto m?\}
\end{array}}
$$

There is of course the possibility that each of the above operational schema will have no effect on the membership of $b?$. For example when $m?$ is less or equal to the existing membership of $b?$, then

$$(fset \text{ or } \{b? \mapsto m?\})(b?)$$
$$= max\{fset(b?), m?\}$$
$$= fset(b?)$$

Similarly when $1 - m?$ is greater or equal to the existing membership of $b?$, then

$$(fset \overset{\mathcal{F}}{\setminus} \{b? \mapsto m?\})(b?)$$
$$= (fset \text{ and not } (\{b? \mapsto m?\}))(b?)$$
$$= (fset \text{ and } \{b? \mapsto (1 - m?)\})(b?)$$
$$= min\{fset(b?), (1 - m?)\}$$
$$= fset(b?)$$

Clearly the specifier needs to recognise this when specifying operational schema of this type.

The specifier could also choose to model a crisp concept as a set of type $\mathbb{P}\, basicType$ rather than of type $\mathcal{C}\, basicType$. Although this is contrary to the philosophy of the toolkit, in practice it is permitted. For example the schema

$$
\boxed{\begin{array}{l}
\underline{\;state\;} \\
cset : \mathbb{P}\, basicType \\
fset : \mathcal{F}\, basicType
\end{array}}
$$

declares two sets, one a power set and the other a fuzzy set within the same specification. However if we wish to include a state constraint or to form a union or intersection between these two sets then we would need to apply the extended set function $(\mathcal{P})$ to $cset$ to ensure type consistency. As the extended set function transforms a set of type $\mathbb{P}\, basicType$ to one of type $\mathcal{C}\, basicType$, it makes more sense to simply model the crisp concept as a set of type $\mathcal{C}\, basicType$ in the first place. Thus

$$
\boxed{\begin{array}{l}
\underline{\;state\;} \\
cset : \mathcal{C}\, basicType \\
fset : \mathcal{F}\, basicType
\end{array}}
$$

is the preferable declaration.

2.5.2   Initialisation to ∅ rather than empty

The specifier could choose to initialise a set of type $\mathcal{F}basicType$ to the empty set rather than to the empty fuzzy set i.e.

```
┌─ INIT ──────────────────────────────────────
│  state
│ ────────────────
│  fset = ∅
└─────────────────────────────────────────────
```

The schema *setAdjust* is unchanged with the fuzzy set *fset* remaining a partial function. If $b?$ is already in the domain of *fset* then the existing membership of $b?$ is replaced by $m?$, otherwise an extra maplet, $b? \mapsto m?$, is added to the set.

```
┌─ setAdjust ─────────────────────────────────
│  Δstate
│  b? : basicType
│  m? : M
│ ────────────────
│  fset' = fset ⊕ {b? ↦ m?}
└─────────────────────────────────────────────
```

However this choice could lead to confusion in the interpretation of zero set membership. Is $b? \notin \text{dom } fset$ to be interpreted differently from $b? \mapsto 0 \in fset$? Or does it mean the same thing? In other words should there be a one-to-one correspondence between the concepts of zero membership and non-membership when applied to fuzzy sets?. This correspondence is certainly implied by the motivation for the development of the toolkit in the first place. One of stated aims for the toolkit development was to extend the set notation to explicitly show the degree of membership of each reference set element. This includes membership values of zero as well as those greater than zero. In this context $b? \notin \text{dom } fset$ is interpreted as indicating that $b?$ is not a member of the reference set on which the fuzzy set is defined rather than not being a member of the fuzzy set itself. Initialisation of a fuzzy set to the empty fuzzy set avoids this confusion by ensuring that all reference elements have a membership value of zero and are thus interpreted as not being a member of the fuzzy set.

There may also be confusion when dealing with the concept of the 'removal' of an element from a fuzzy set. Is the removal of $b?$ from a set of type $\mathcal{F}basicType$ to be interpreted as an adjustment of membership to zero or is it that $b? \mapsto m?$ is removed altogether i.e. $b?$ is no longer in dom *fset*? If this is not clarified then a situation could arise where an operational schema to 'remove' an element from a fuzzy set may in fact lead to the addition of a new maplet to the set. For example

---

remove
$\Delta state$
$b? : basicType$

$fset' = fset \oplus \{b? \mapsto 0\}$

The intent of the schema is that the membership of $b?$ is to be adjusted to zero. Presumably this is being interpreted as indicating that $b?$ is being removed from $fset$. However in a situation where $b? \notin \operatorname{dom} fset$ the result will be that $b? \mapsto 0$ will be _added_ to $fset$, which is counter intuitive. Obviously this could be overcome by including a pre-condition that ensures that the operation only succeeds when $b? \in \operatorname{dom} fset$ i.e.

---

remove
$\Delta state$
$b? : basicType$

$b? \in \operatorname{dom} fset$
$fset' = fset \oplus \{b? \mapsto 0\}$

However what is really happening now is that the schema is checking whether $b?$ is in the reference set for the fuzzy set, rather than whether it has any membership in the fuzzy set on which the operation is be applied. Remember that this pre-condition will hold when $b? \mapsto 0$ is a member of $fset$ so the operation will succeed even if the element is already 'removed'. Intuitively we would expect the pre-condition to be a constraint on the intent of the operational schema rather than on the definition of the set itself.

The initialisation of a fuzzy set to the empty fuzzy set avoids all of this. It ensures that the membership value for any reference set element is always defined and that removal should result in the element no longer being a member of the fuzzy set. In other words that the membership value is adjusted to zero. This is consistent with the interpretation of zero set membership as indicating non-fuzzy set membership. The meaning of an operational schema to remove element from the fuzzy set becomes quite clear and has the same meaning in the crisp case as we would expect in conventional Z

Using the toolkit   Using conventional Z

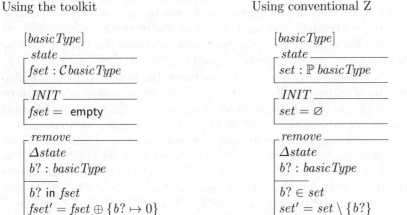

2.5.3   Guidelines for toolkit use

The previous discusion can be summarised as a series of guidelines for toolkit use. It should be remembered that a specification written using the fuzzy logic toolkit is still being written in Z and consequently all aspects of the language and the mathemetical toolkit are available and can be used. Adherence to the following guidelines should ensure that a specification is consistent with the underlying philosophy of the toolkit.

- When developing a specification that includes both fuzzy and crisp concepts, boolean sets should be declared as being of type $\mathcal{C}\,basicType$, rather than $\mathbb{P}\,basicType$.
- Fuzzy (and crisp) sets should be initialised to the empty fuzzy set rather than to the empty set. This is consistent with the semantics of the toolkit when dealing with the concept of set emptiness.
- When writing operational schema that adjusts set membership the function override operator, $\oplus$, should be used. The concepts of adding and removing elements to and from a set in the crisp case is generalised to an adjustment of set membership in the fuzzy case.

2.6   Conclusion

It has been suggested that some system problems are not naturally understood in precise or exact terms. They mainly arise in a humanistic context and are characterised by inherent uncertainty and imprecision. Many involve human decision making and judgement. It has also been suggested that there are advantages to be gained by using formal methods as early as possible in systems development. However the usefulness of existing formal method languages such as Z is limited when applied to these softer problem types. Fuzzy set theory has been suggested as a possible formalism for modelling some of the uncertainty and imprecision typical of these systems. A fuzzy logic toolkit for Z that defines the operators, modifiers and measures necessary for the manipulation of fuzzy sets and relations has been developed. The illustrative examples are designed to show how the toolkit might be applied to the type of system problems of interest. The emphasis has been on those aspects of the problem domain where it is more natural to think of the extent or the degree to which something occurs, or where it is more natural to describe concepts approximately and linguistically rather than in a precise or numeric way. An attempt has been made to focus on the 'grayness' in the problem domain resulting in models that reflect the underlying fuzziness characteristic of these problem types. Specifications written using the toolkit are written in Z and consequently the specifier has available all aspects of the language. The guidelines for toolkit usage have been developed to provide some consistency between a specification written using the toolkit and the underlying ideas that governed the development of the toolkit itself.

The work described in this chapter is ongoing. The toolkit is currently been extended to include concepts such as the closure of a fuzzy relation and a set of more general definitions for a fuzzy number. Consideration is also being given to the inclusion of a fuzzy function override operator and the more general concept of a type 2 fuzzy set in future versions of the toolkit. It is anticipated that each addition to the toolkit will be supported by an appropriate illustrative example. Currently a specification for a simple fuzzy expert system is being developed which is intended to illustrate the use of the truncation and scaling operators for a fuzzy set, and the fuzzy cartesian product operator.

References

1. G.M. Allinger, S.L. Feinzig, and E.A. Janak. Fuzzy Sets and Personnel Selection: Discussion and an Application. *Journal of Occupational and Organizational Psychology*, 66:162–169, 1993.
2. G. Bojadziev and M. Bojadziev. *Fuzzy Sets, Fuzzy Logic, Applications*. World Scientific, Singapore, 1995.
3. Bruce G. Buchanan and Edward H. Shortliffe, editors. *Rule-Based Expert Systems: The MYCIN Experiments of the Stanford Heuristic Programming Project*. Addison-Wesley Publishing Company, Reading, Massachusetts, 1985.
4. E. Castillo and E. Alvarez. *Expert Systems: Uncertainty and Learning*. Elsevier Science Publishers Ltd, Barking, Essex, 1991.
5. Concepcion San Luis Costa, Pedro Prieto Maranon, and Juan A. Hernandez Cabrera. Application of diffuse measurement to the evaluation of psychological structures. *Quality & Quantity*, 28:305–313, 1994.
6. E. Cox. *The Fuzzy Systems Handbook*. AP Professional - Harcourt Brace & Company, Boston, 1994.

7. D. Dubois and H. Padre. *Fuzzy Sets and Systems: Theory and Applications.* Academic Press, Inc, 1980.

8. D. Dubois and H. Prade. *Possibility Theory - An Approach to Computerised Processing of Uncertainty.* Plenum Press, New York, 1988.

9. Michael W. Eysenck and Mark T. Keane. *Cognitive Psychology.* Lawrence Erlbaum Associates, Hove, 1995.

10. I. Graham. Fuzzy Logic in Commercial Expert Systems - Results and Prospects. *Fuzzy Sets and Systems*, 40:451–472, 1991.

11. Susan Haack. *Deviant Logic, Fuzzy Logic — Beyond Formalism.* University of Chicago Press, Chicago, second edition, 1996.

12. B. Hesketh, R. Prior, M. Gleitzman, and T. Hesketh. Practical Applications and Psychometric Evaluation of a Computerised Fuzzy Graphic Rating Scale. In T. Zetenyi, editor, *Fuzzy Sets in Psychology*, pages 425–454. Elsevier Science Publishers B.V, North-Holland, 1988.

13. B. Hesketh, R. Pryor, and M. Gleitzman. Fuzzy Logic: Toward Measuring Gottfredson's Concept of Occupational Social Space. *Journal of Counselling Psychology*, 36(1):103–109, 1989.

14. Peter Jackson. *Introduction to Expert Systems.* Addison Wesley Longman, Harlow, England, third edition, 1999.

15. J. Jacky. *The Way of Z: Practical Programming with Formal Methods.* Cambridge University Press, Cambridge, 1997.

16. Jyh-Shing Roger Jang, Chuen-Tsai Sun, and Eiji Mizutani. *Neuro-Fuzzy and Soft Computing - A Computational Approach to Learning and Machine Intelligence.* Prentice-Hall, Inc., New Jersey, 1997.

17. J. Kacprzyk, M. Fedrizzi, and H. Nurmi. Fuzzy Logic with Linguistic Quantifiers in Group Decision Making. In R.R. Yager and L.A. Zadeh, editors, *An Introduction to Fuzzy Logic Applications in Intelligent Systems*, pages 263–280. Kluwer Academic, 1992.

18. J. Kacprzyk, M. Fedrizzi, and H. Nurmi. Soft Degrees of Consensus under Fuzzy Preferences and Fuzzy Majorities. In Janusz Kacprzyk, Hannu Nurmi, and Mario Fedrizzi, editors, *Consensus under Fuzziness*, pages 56–81. Kluwer Academic, 1997.

19. A. Kaufmann. *Introduction to the theory of Fuzzy Subsets*, volume 1 - Fundamental Theoretical Elements. Academic Press, London, 1975.

20. Joseph F. Kess. *Psycholinguistics - Psychology, Linguistics and the study of natural language.* John Benjamin, Amsterdam/Philadelphia, 1992.

21. G. J. Klir and D. Harmanec. Types and Measures of Uncertainty. In Janusz Kacprzyk, Hannu Nurmi, and Mario Fedrizzi, editors, *Consensus under Fuzziness*, pages 29–51. Kluwer Academic, 1997.

22. B. Kosko. Fuzziness vs. Probability. *Int. J. General Systems*, 17:211–240, 1990.

23. B. Kosko. *Neural Networks and Fuzzy Systems.* Prentice-Hall, New Jersey, 1992.

24. R. Kruse, J. Gebhardt, and F. Klawonn. *Foundations of Fuzzy Systems.* John Wiley & Sons, Chichester, 1994.

25. J. Lee and J. Yen. Specifying Soft Requirements of Knowledge-Based Systems. In R.R Yager and L.A. Zadeh, editors, *Fuzzy Sets, Neural Networks, and Soft Computing*, pages 285–295. Van Nostrand Reinhold, New York, 1994.

26. R. Lowen. *Fuzzy Set Theory: Basic Concepts, Techniques and Bibliography.* Kluwer Academic, Dordrecht, 1996.

27. A. De Luca and S. Termini. A Definition of a Nonprobablistic Entropy in the Setting of Fuzzy Set Theory. *Information and Control*, 20:301–312, 1972.
28. C. Matthews and P. A. Swatman. Fuzzy Concepts and Formal methods : Some Illustrative Examples. Technical Report 1999:37, School of Management Information Systems, Deakin University, 1999.
29. C. Matthews and P. A. Swatman. Fuzzy Z - The Extended Notation, revision 2 (2000). Technical Report 1999:38, School of Management Information Systems, Deakin University, 1999.
30. C. Matthews and P. A. Swatman. Fuzzy Concepts and Formal Methods : A Fuzzy Logic Toolkit for Z. In *ZB2000: Formal Specification and Development in Z and B, First International Conference of B and Z Users York, UK, 29 Aug. to 2 Sept. 2000*, Lecture Notes in Computer Science. Springer-Verlag, 2000.
31. Douglas L. Medin and Brian H. Ross. *Cognitive Psychology*. Harcourt Brace Jovanovich, Inc., Orlando, 1992.
32. S. E. Newstead. Quantifiers as Fuzzy Concepts. In T. Zetenyi, editor, *Fuzzy Sets in Psychology*, pages 51–72. Elsevier Science Publishers B.V, North-Holland, 1988.
33. J. Nichols. Z notation — Version 1.3. Technical report, ISO, June 1998.
34. Zdzislaw Pawlak. An inquiry into anatomy of conflicts. *Journal of Information Sciences*, 109:65 – 78, 1998.
35. B. Potter, J. Sinclair, and D. Till. *An Introduction to Formal Specification and Z*. Prentice Hall International Series in Computer Science, Hemel Hempstead, second edition, 1996.
36. Roger S. Pressman. *Software Engineering A Practioner's Approach*. McGraw-Hill, International Edition, fourth edition, 1997.
37. H. Saiedian. Formal Methods in Information Systems Engineering. In R. H Thayer and M. Dorfman, editors, *Software Requirements Engineering*, pages 336–349. IEEE Computer Society Press, second edition, 1997.
38. K.J. Schmucker. *Fuzzy Sets, Natural Language Computations, and Risk Analysis*. Computer Science Press, Rockville, 1984.
39. M. Smithson. *Fuzzy Set Analysis for Behavioral and Social Sciences*. Springer-Verlag, New York, 1987.
40. M. Smithson. *Ignorance and Uncertainty - Emerging Paradigms*. Springer-Verlag, New York, 1988.
41. J.M Spivey. *The Z Notation: A Reference Manual*. Prentice Hall International Series in Computer Science, Hemel Hempstead, second edition, 1992.
42. I. Toyn. Innovations in the Notation of Standard Z. In J. P Bowen, A. Fett, and M. G. Hinchey, editors, *ZUM '98: The Z Formal Specification Notation*, Lecture Notes in Computer Science. Springer-Verlag, 1998.
43. G. Viot. Fuzzy Logic: Concepts to Constructs. *AI Expert*, 8(11):26–33, November 1993.
44. M. Viswanathan, M. Bergen, S. Dutta, and T. Childers. Does a Single Response Category in a Scale Completely Capture a Response? *Psychology and Marketing*, 13(5):457–479, 1996.
45. J. M. Wing. A Specifier's Introduction to Formal Methods. *IEEE Computer*, 23(9):8–24, 1990.
46. Ronald R. Yager. Fuzzy logic in the formulation of decision functions from linguistic specifications. *Kybernetes*, 25(4):119–130, 1996.
47. Y. Y. Yao. A comparative study of rough sets and fuzzy sets. *Journal of Information Sciences*, 109:227–242, 1998.

48. L. A. Zadeh. Fuzzy Sets. *Information and Control*, 8:338–353, 1965.
49. L. A. Zadeh. The Concept of a Linguistic Variable and its Application to Approximate Reasoning I. *Information Sciences*, 8(4):199–249, 1975.
50. L. A. Zadeh. Fuzzy Logic. *IEEE Computer*, 21(4):83–92, April 1988.
51. L. A. Zadeh. Knowledge Representation in Fuzzy Logic. In R.R. Yager and L.A. Zadeh, editors, *An Introduction to Fuzzy Logic Applications in Intelligent Systems*, pages 1–25. Kluwer Academic, 1992.
52. Qiao Zhang. Fuzziness – vagueness – generality – ambiguity. *Journal of Pragmatics*, 29:13–31, 1998.
53. A.C. Zimmer. A Common Framework for Colloquil Quantifiers and Probability Terms. In T. Zetenyi, editor, *Fuzzy Sets in Psychology*, pages 73–89. Elsevier Science Publishers B.V, North-Holland, 1988.
54. R. Zwick, D. V. Budescu, and T. S. Wallsten. An empirical study of the integration of linguistic probabilities. In T. Zetenyi, editor, *Fuzzy Sets in Psychology*, pages 91–125. Elsevier Science Publishers B.V, North-Holland, 1988.

3 Trade-off Requirement Engineering

Jonathan Lee[1], Jong-Yih Kuo[2], Nien-Lin Hsueh[3], and Yong-Yi Fanjiang[1]

[1] Software Engineering Lab.
  Department of Computer Science and Information Engineering
  National Central University
  Chungli, Taiwan
  {yjlee,yyfanj}@selab.csie.ncu.edu.tw
[2] Computer Science and Information Engineering
  Fu-Jen Univesity
  HsinChung, Taiwan
  yjkuo@csie.fju.edu.tw
[3] Department of Information Management
  Shu-Te University
  Kaoshiung, Taiwan
  nien@mail.stu.edu.tw

3.1 Introduction

A major challenge in requirements engineering of knowledge-based systems is that the requirements to be captured are imprecise in nature and usually conflicting with each other [25,37,48]. Balzer et al. have argued that informality is an inevitable (and ultimately desirable) feature of the specification process [2]. Similar ideas are also advocated by other researchers such as Borgida [5], Feather [16], Fickas [12], Waters [36], Niskier [33] and their colleagues. Borgida et al. have further elaborated that a good requirement modeling approach should take the problem of describing nature kinds into account, which usually runs the risk of being *vague* and subject to *contradiction* [5].

However, most of the existing work on requirements modeling are limited in dealing with this problem. Traditional requirements modeling (formal or informal) either requires the requirements be stated precisely or completely excludes this problem out of the scope of the modeling activity (e.g. see [13] for a survey). Knowledge-based software engineering indirectly addresses problems caused by the vagueness in the specification process by converting domain-specific informal requirements into formal ones [21,30].

In this chapter, we introduce a integrated approach for acquiring heterogeneous requirement in the elicitation phase, modeling vague requirements as fuzzy object model in the modeling phase, and analyzing requirement in the analysis phase (see Fig. 3.1). In the elicitation phase, we argue that there are various kinds of information in the informal requirements (i.e. heterogeneous), and that to appropriately model the requirements, we propose a goal-driven-use-case(called GDUC) approach for eliciting and structure requirements [27,28]. In GDUC, a faceted classification scheme is proposed for identifying goals from domain descriptions and system requirements. Each

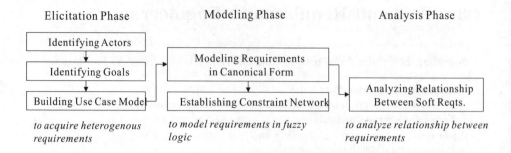

Fig. 3.1. An Overview of Trade-Off Requirement Engineering

goal can be classified under three facets we have identified: competence, view and content. More detailed requirements (represented as use case) can be acquired by asking how the goals are achieved, maintained or optimized.

We then formulate soft functional requirements based on fuzzy logic [49] and an extension of the notion of soft conditions in TBSM [46,26] to alleviate the difficulties in representing vague requirements. More specifically, the soft functional requirement is represented using the canonical form in test-score semantics [50]. The trade-offs among conflicting requirements are analyzed by identified the relationship between requirements which could be either conflicting, irrelevant, cooperating, or independent. A requirement hierarchy is established based on the notion of criticality and cooperating degree. A parameterized aggregation operator, fuzzy and / or, is selected to combine individual requirements. A feasible overall requirement can be obtained through the aggregation of individual requirements based on the requirements hierarchy.

This chapter is organized as follows. We first introduce our goal-driven requirement elicitation approach in the next section. The notion and formalization of soft functional requirements is discussed in section 3.3. The proposed approach for analyzing soft functional requirements is presented in section 3.4. Related work is outlined in Section 3.5. Finally, we summarize the potential benefits of the proposed approach and our future research plan in section 3.6.

3.2   Requirements Elicitation

As was observed by Blum that informal requirements are heterogeneous, and can be organized in a variety of ways [4]. In this section, a goal-driven use case approach is proposed to facilitate the elicitation of informal requirements [27].

Goal-driven use cases is an approach for requirements engineers to elicit and structure users requirements, and to analyze and evaluate relationships between requirements. There are three steps to construct use cases: (1) identify actors by investigating all possible types of user that interact with the system directly; (2) identify goals based on a faceted classification scheme; and (3) build use case models.

Identifying Actors. An actor is an outside entity that interacts directly with a system, which may be a person or a quasi-autonomous object, such as machines, computer tasks, and other systems. More precisely, an actor is a role played by such an entity. For example, the meeting scheduler system is mainly designed for an initiator to organize a meeting schedule, and therefore, *initiator* is marked as an actor.

Identifying Goals. A faceted classification scheme is proposed for identifying goals from domain descriptions and system requirements. Each goal can be classified under three facets we have identified: competence, view and content. The facet of competence is related to whether a goal is completely satisfied or only to a degree. A *rigid* goal describes a minimum requirement for a target system, which is required to be satisfied utterly. A *soft* goal describes a desirable property for a target system, and can be satisfied to a degree. For example, if a meeting schedule is convenient for all attendants, the goal *MaxConvenienceSchedule* is defined to be satisfied completely. However, if the schedule is only convenient to only some of the attendants, the goal is said to be satisfied to a degree. A soft goal is related to a rigid one in the sense that the existence of the soft goal is dependent on the rigid one.

The facet of view concerns whether a goal is actor-specific or system-specific. Actor-specific goals are objectives of an actor in using a system; meanwhile, system-specific goals are requirements on services that the system provides. For example, through examining the system description, we have found out that the initiator has three objectives in using the meeting scheduler system: (1) to create a meeting, (2) to make the meeting schedule as convenient as possible for the participants, and (3) to maximize the number of participants for the meeting. Therefore, three actor-specific goals can be identified: *MeetingRequestSatisfied*, *MaxNumberOfParticipants* and *MaxConvenienceSchedule*. On the other hand, a system-specific goal takes into consideration "what kinds of properties the system should have in supporting services to *all* users?", or "what are the requirements on the services for the system to provide?". In our example, to construct a meeting is an objective of an initiator, but to accommodate a more important meeting is a requirement of the system. Therefore, a system-specific goal — *SupportFlexibility*, is identified.

Usually, requirements can be classified into functional and nonfunctional requirements based on their content [24]. The construction of functional requirements involves modeling the relevant internal states and behavior of both the component and its environment. Nonfunctional requirements usually define the constraints that the product needs to satisfy. Therefore, a goal can be further distinguished based on its content, that is, a goal can be either related to the functional aspects of a system or associated with the nonfunctional aspect of the system[1]. A functional goal can be achieved by performing a sequence of operations. A nonfunctional goal is defined as constraints to qualify its related functional goal. In our example, creating a meeting is accomplished by a sequence of operations, therefore the goal *MeetingRequestSatisfied* is defined as a functional goal. On the other hand, goals like *MaxNumberOfParticipants* and *MaxConvenienceSchedule* can be viewed as constraints for a schedule to satisfy, which are identified as nonfunctional goals.

Building Use Case Models. To structure a use case and its extensions, we extend the work by Cockburn [8] by considering several different types of goal. Essentially, each use case is viewed as a process that can be associated with a goal to be achieved, optimized, or maintained by the use case. To start with, we first consider original use cases to guarantee that the target system will be at least adapted to the minimum requirements. Each original use case in our approach is associated with an actor to describe the process to achieve an *original goal* which is rigid, actor-specific and functional. Building original use cases by investigating all original goals will make the use case model satisfy at least all actors' rigid and functional goals.

The basic course in an original use case is the simplest course, the one in which the goal is delivered without any difficulty. The alternative course encompasses the recovery course and/or the failure one. The recovery course describe the process to recover the original goal, whereas the failure one describes what to do if the original goal is not recoverable.

In our example, the use case *plan a meeting* covers the case for an initiator to achieve the goal *MeetingRequestSatisfied* (see Fig. 3.2) which is rigid, actor-specific and functional. The use case starts when an initiator issues a meeting request to the system, and lasts until a meeting schedule is generated or canceled. It is the basic course that forms the foundation when specifying a use case and this should be described first. The use case has several alternative courses that may change its flow. An example of this is different ways of recovering the goal *MeetingRequestSatisfied* when there exists a strong conflict in a schedule.

Original use cases are designed to satisfy original goals for modeling users minimum requirements. To extend the model to take into account different types of goal, extension use cases are created. Situations about when to create extension use cases are fully discussed below:

- To optimize or maintain a soft goal. By achieving a rigid goal, all its related soft goals can also be satisfied to some extent. To optimize or maintain the soft goals, extension use cases are created. Therefore, the

---

[1] Similar ideas are advocated in [1], where achievement and maintenance goals map to actions and nonfunctional requirements, respectively.

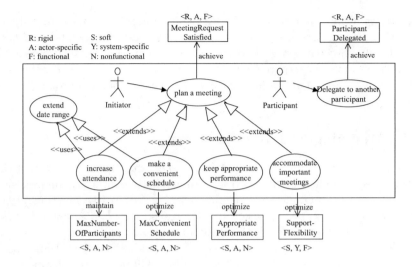

Fig. 3.2. A Goal-Driven Use Case Model for Meeting Scheduler System

basic course in an extension is to optimize (or maintain) its soft goal, whereas an alternative course describes what to do if it fails to optimize (or maintain) the goal. In our example, to satisfy the rigid goal *MeetingRequestSatisfied* does not guarantee that the meeting is convenient for all participants. To make the schedule as convenient as possible, the extension *make a convenient schedule* is created. If the constraints in the basic course are not satisfied, the alternative course is to recover the optimization of the soft goal, for example, to extend the date range, or to ask participants to add dates to their preference sets.

- To achieve a system-specific goal. An extension use case may be created to achieve a system-specific goal. Referring to our example, the original use case *plan a meeting* describes the process to create a meeting from a personal view (the view of the actor *initiator*). The extension use case *accommodate important meetings* extends it to take all initiator into account, that is, to achieve a system-specific goal *SupportFlexibility*.

- To achieve a nonfunctional goal. To extend a use case model to capture nonfunctional requirements, extension use cases are added to achieve a nonfunctional goal. In this case, an extension use case serves as a constraint to qualify its original use case. In our example, the original use case *plan a meeting* is a direct course to create a meeting, several constraints (may be rigid or soft) on a meeting are ignored: *AppropriatePerformance*, *MaxNumerOfParticipants* and *MaxConvenienceSchedule*. The basic course of *make a convenient schedule* indicates the soft constraints on a meeting schedule. If the constraints are not satisfied, the alternative course is to recover the optimization of the soft goal.

Identifying Common Ground. Use case description is process to help to understanding the requirements. By inspecting the requirements, analyzers can identify the common ground for different requirements. In the meeting scheduler system, the requirements of *increase attendance, make a convenient schedule* and *keep appropriate performance* have the common ground – date range.

3.3   Modeling Soft Requirements

In this section, we propose the use of *soft requirements*, an extension of the notion of soft conditions in TBSM [47], to explicitly capture the imprecision of functional and nonfunctional requirements. In TBSM, the functionality of a task is specified by properties of its state-transition $<b, a>$, where b is the state before the task, and a is the state after invoking the task.

The functional requirement of a task can thus be specified using a pair $<precondition, postcondition>$. In the traditional approach to software specification, the precondition and the postcondition describe properties that should be held by the states b and a. These conditions specify a rigid functionality because whether they are satisfied by a state is a black-and-white situation. Rigid functional requirements can be formally defined as follows:

Definition 1. *(Rigid Functional Requirement)*:
A rigid functional requirement of a task T is a pair of formula $< \varphi_1, \varphi_2 >$ where φ_1 is a precondition and φ_2 is a postcondition, such that for every $<b, a> \in$ T,

$$hold(\varphi_1, b) \implies hold(\varphi_2, a)$$

where "$hold(\varphi, b)$" is a function that returns either 1 (true) or 0 (false) in a state a, and $\implies$ denotes logic implication operator.

However, as was mentioned earlier, imprecision often is inevitable in the functional requirements of complex software systems. We propose the use of soft functional requirements to directly express requirements that are elastic in nature. A soft functional requirement describes state properties that can be satisfied to a degree. By generalizing the definition of rigid functional requirements, we arrive at the following formal definition of soft functional requirements:

Definition 2. *(Soft Functional Requirement)*:
A soft functional requirement of a task T is a pair of formula $< \varphi_1, \varphi_2 >$ where φ_1 is T's precondition and φ_2 is T's postcondition, such that for every $<b, a> \in$ T,

$$fhold(\varphi_1, b) \overset{f}{\implies} fhold(\varphi_2, a)$$

where "$fhold(\varphi, b)$" is a function that returns the degree to which a formula φ is true in state s, and $\overset{f}{\implies}$ denotes implication operator in fuzzy logic.

The function *fhold* is a generalization of the *hold* predicate in situation calculus [32], which states properties that are true in a given state. Note that a soft functional requirement degenerates to a rigid functional requirement when $fhold(\varphi, s)$ returns either 1 (true) or 0 (false). Therefore, a soft functional requirement is a generalization of a rigid functional requirement.

The soft conditions can be represented using the notion of *canonical form* in Zadeh's test-score semantics [50]. A basic idea underlies test-score semantics is that a proposition p in a natural language may be viewed as a collection of *elastic constraints*, $C_1, \ldots, C_k$, which restricts the values of a collection of variables $X = (X_1, \ldots, X_n)$. In fuzzy logic, this is accomplished by representing p in the canonical form

$$p \rightarrow X \ is \ A$$

in which A is a fuzzy predicate or, equivalently, an n-ary fuzzy relation in U, where $U = U_1 \times U_2 \times \cdots \times U_n$ and U_i, $i = 1, \ldots, n$, is the domain of X_i. The canonical form of p implies that the possibility distribution of X is equal to $A \ \pi_X = A$ which in turns implies that $Poss\{X = u\} = \mu_A(u)$, $u \in U$, where μ_A is the membership function of A and $Poss\{X = u\}$ is the possibility that X takes u as its value [49]. It is in this sense that A, acting as an elastic constraint on X, restricts the possible values which X can take in U.

Theorem 1. *Let p be a proposition in its canonical form, X is A, X is a state variable in state s, and u_i is the value of X in s. Then*

$$fhold(p, s) = \mu_A(u_i)$$

This theorem can be trivially proved by applying test-score semantics to situation calculus.

3.4   Analyzing Soft Requirements

There are in general three steps involved in the proposed trade-offs analysis technique for soft requirements: (1) to examine the conflicting and cooperating degrees for any two individual soft requirements; (2) to identify the relationships between soft requirements based on the conflicting and cooperating degrees; and (3) to aggregate soft requirements for a feasible overall requirement.

3.4.1   Defining Conflicting and Cooperating Degrees

Intuitively, two requirements are conflicting with each other if an increase in the degree to which one requirement is satisfied often decreases the degree to which another requirement is satisfied, that is, the *fhold* decreases between the two after states (called a conflicting after state pair). On the other hand, two requirements are said to cooperate with each other if an increase (or decrease) in the degree to which one requirement is satisfied often increases (or decrease) the degree to which another requirement is satisfied, that is, the *fhold* increases between the two after states (called a cooperating after state pair). Note that the third possibility is that the *fhold* remains unchanged between the two after states, which is called an irrelevant after state pair. We formally define conflicting, cooperating and irrelevant after state pairs below, and conflicting and cooperating degrees wrt a given before state, which is followed by a formal definition of average conflicting and cooperating degrees.

Definition 3. *(Conflicting, Cooperating and Irrelevant After State Pairs):* Assume that $R_1 =< \varphi_{11}, \varphi_{12} >$, $R_2 =< \varphi_{21}, \varphi_{22} >$ are two requirements, and that A_b is a set of common after state pairs wrt a given before state b. A set of conflicting after state pairs, denoted as CF_b, is defined as

$$\{(a_i, a_j) \mid (fhold(\varphi_{12}, a_i) - fhold(\varphi_{12}, a_j)) \times (fhold(\varphi_{22}, a_i) - fhold(\varphi_{22}, a_j)) < 0, and\ a_i, a_j \in A_b\}$$

A set of cooperating after state pairs, denoted as CP_b, is defined as

$$\{(a_i, a_j) \mid (fhold(\varphi_{12}, a_i) - fhold(\varphi_{12}, a_j)) \times (fhold(\varphi_{22}, a_i) - fhold(\varphi_{22}, a_j)) > 0, and\ a_i, a_j \in A_b\}$$

A set of irrelevant after state pairs, denoted as IR_b, is defined as

$$\{(a_i, a_j) \mid (fhold(\varphi_{12}, a_i) - fhold(\varphi_{12}, a_j)) \times (fhold(\varphi_{22}, a_i) - fhold(\varphi_{22}, a_j)) = 0, and\ a_i, a_j \in A_b\}$$

Hence, a set of common after state pairs, A_b, can be divided into three classes, in such a way that
$A_b = CF_b \cup CP_b \cup IR_b$ and $CF_b \cap CP_b \cap IR_b = \phi$.

Definition 4. *(Conflicting and Cooperating Degrees):*

Assume that $R_1 =< \varphi_{11}, \varphi_{12} >$, $R_2 =< \varphi_{21}, \varphi_{22} >$ are two requirements, A_b is a set of common after state pairs wrt a given before state b, and that CF_b and CP_b denote conflicting and cooperating after state pairs wrt b, respectively.
The conflicting degree between two requirements, R_1 and R_2, wrt b, is defined as:

$$cf_b(R_1, R_2) = \frac{\sum_{a_i, a_j \in CF_b} (|fhold(\varphi_{12}, a_i) - fhold(\varphi_{12}, a_j)| + |fhold(\varphi_{22}, a_i) - fhold(\varphi_{22}, a_j)|)}{\sum_{a_h, a_k \in A_b} (|fhold(\varphi_{12}, a_h) - fhold(\varphi_{12}, a_k)| + |fhold(\varphi_{22}, a_h) - fhold\varphi_{22}, a_k)|)}$$

The cooperating degree between two requirements, R_1 and R_2, wrt b, is defined as:

$$cp_b(R_1, R_2) = \frac{\sum_{a_i, a_j \in CP_b} (|fhold(\varphi_{12}, a_i) - fhold(\varphi_{12}, a_j)| + |fhold(\varphi_{22}, a_i) - fhold(\varphi_{22}, a_j)|)}{\sum_{a_h, a_k \in A_b} (|fhold(\varphi_{12}, a_h) - fhold(\varphi_{12}, a_k)| + |fhold(\varphi_{22}, a_h) - fhold(\varphi_{22}, a_k)|)}$$

Based on Definition 3, we know that it may has many different state after performing a task T. A requirement can be specified using "state change" from before state through after state. A soft requirement of the target system may have many before states and after states, so the relationship of cooperating and conflicting between soft requirements must consider all possible before states. We are ready to introduce average cooperating and conflicting degree.

Definition 5. *(Average Conflicting and Cooperating Degrees)*:
Assume that $R_1 = < \varphi_{11}, \varphi_{12} >$, $R_2 = < \varphi_{21}, \varphi_{22} >$ are two requirements, A_b is a set of common after state pairs wrt a given before state b, B_{R_1} and B_{R_2} are sets of all possible before states of R_1 and R_2. The average conflicting degree between two requirements, R_1 and R_2, wrt b, is defined as:

$$cf(R_1, R_2) = \frac{\sum_{b_k \in B_{R_1} \cap B_{R_2}} cf(R_1, R_2)}{||B_{R_1} \cap B_{R_2}||}$$

The average cooperating degree between two requirements, R_1 and R_2, wrt b, is defined as:

$$cp(R_1, R_2) = \frac{\sum_{b_k \in B_{R_1} \cap B_{R_2}} cp(R_1, R_2)}{||B_{R_1} \cap B_{R_2}||}$$

where $|| B_{R_1} \cap B_{R_2} ||$ is the cardinality of intersection between B_{R_1} and B_{R_2}.

We have further distinguished positive cooperating degrees from negative ones. A positive cooperating degree means that if an increase in the degree to which one requirement is satisfied often increases the degree to which another requirement is satisfied. Otherwise, it is called negative cooperating degrees.

3.4.2   Relationships between Soft Requirements

The relationships among soft requirements are crucial for adequately interpreting the intended meaning of an overall feasible requirement, because they reflect the structure of interaction among the soft requirements and represent users' pros and cons of the requirements. Together with information about the criticality of soft requirements, the relationships among soft requirements can serve as a guideline for requirements aggregations. Similar ideas can also be found in multi-objectives decision making [19,7].

In the proposed framework, the relationship between any two soft requirements, say R_i and R_j, can be classified under four categories: independent, irrelevant, conflicting and cooperating. R_i and R_j are said to be *independent* if there is no common after state shared by the requirements, that is, $A_{R_1} \cap A_{R_2} = \varnothing$. However, all after state pairs are irrelevant pairs (i.e. $cp = cf = 0$) are said to be irrelevant, and the conflicting and cooperating degrees are equivalent to 0.5, R_i and R_j are said to be *counterbalance*.

To further refine conflicting and cooperating relationships, we have identified three sub-categories: strong, moderate and weak. A relationship is said to be conflicting if the conflicting degree between R_i and R_j is greater than the cooperating degree. On the other hand, if the cooperating degree is greater than the conflicting degree, then R_i cooperates with R_j. In the case that there is only conflicting after state pairs (i.e. $cp = 0$), R_i is *strongly conflicting* with R_j. Similarly, if there is only cooperating after state pairs (i.e. $cf = 0$), two requirements are *strongly cooperating* with each other.

Table 3.1. Relationships between Soft Requirements

relation / condition	conflicting			irrelevant	counterbalance	cooperating			independent
	strong	moderate	weak			strong	moderate	weak	
cf-cp	cp=0	cf-cp>0	cf-cp>0	cf=cp	cf=cp=0.5	cf=0	cf-cp<0	cf-cp<0	$A_{R1} \cap A_{R2} = \phi$
cp+cf	cf=1	cf+cp=1	cf+cp<1	cf+cp=0	cf+cp=1	cp=1	cp+cf=1	cf+cp<1	
cp: cooperating degree cf: conflicting degree									

The co-existence of conflicting, cooperating and irrelevant after state pairs (i.e. $cp + cf < 1$) usually drops either the conflicting degree or the cooperating degree further compared with the existence of only conflicting and cooperating after state pairs (i.e. $cp + cf = 1$). The former is called *weak conflicting* or *weak cooperating*, while the later is called *moderate*. The relationships between soft requirements are summarized in Table 3.1.

3.4.3   Hierarchical Aggregation of Requirements

Having decomposed user's requirements into different individual ones, it is then necessary to achieve some level of integration between those individual requirements. In general, there are two issues needed to be addressed in using aggregation operators for the integration of individual requirements.

- The variety of aggregation operators could make it difficult to determine which one to use in a specific application [22]. Zimmermann [51] has outlined eight general criteria: axiomatic strength, empirical fit, adaptability, numerical efficiency, compensation, range of compensation, aggregating behavior, and required scale level of membership functions. Furthermore, Yen et al. [48] have summarized several criteria for selecting an appropriate aggregation operator in the context of requirements engineering: (1) intended relationship between requirements, (2) feasible combined requirements, (3) higher satisfiable realization, and (4) requirements with criticalities. However, most of the existing approaches only considered one or two of the criteria in the analysis, for example, requirements with criticalities in [46], intended relationships between requirements in [48] or between objectives in [19].

- The averaging operator is symmetrical, monotonic, commutative and idempotency, but the property of associativity is not available. The lack of associativity with respect to any averaging operator raises some important issues of how to extend the operator. Several researchers such as Yager [45] and Cutello & Motero [11] have proposed different imperatives in holding the definition together as elements are added to an aggregation.

To alleviate the above problems, we propose an extension of the hierarchical aggregation structure advocated in [44], where requirements in each disjunct and conjunct are expanded to form a requirements hierarchy. In the proposed framework, we not only explore all the criteria summarized by Yen et al. in [48], but also relax the assumption by taking the consideration that requirements from users are usually described using either *and*, or *or* natural language connectives, or both. The steps in establishing a hierarchical structure for requirements aggregation are discussed below.

Convert requirements into DNF. To take these connectives into account, we have proposed the use of disjunctive normal form (DNF) to obtain a uniform representation of the requirements. Requirements in DNF can then be arranged based on an extension of the notion of the hierarchical aggregation structure advocated in [52,44,9,11], where requirements in each disjunct and conjunct are expanded to form a requirements hierarchy. A requirements hierarchy can be established based on the notion of the requirements criticality and the positive cooperative degree. In fact, the requirements may carry different weights reflecting their degrees of criticality, where a weight is a non-negative real number. We have adopted Saaty's pairwise comparison approach to the assignment of weights to requirements [40]. That is, the relative weights of each requirement pair are used to form a reciprocal matrix, and the absolute weight of each requirement is obtained from the normalized eigenvector using eigenvalue method. A requirement R can thus be represented by a triple: $< w_R, \mu_A(x), R >$, where w_R denotes the criticality associated with the requirement R, and $\mu_A(x)$ is derived from Definition 1, $x \in X$.

Establish a requirements hierarchy. Assume that requirements are either connected by the conjunction or by the disjunction connective and that the hierarchy is from level 0 (the top level) to level n.

Algorithm 1 *(Establish a Requirements Hierarchy)*

1. *Top-down:*
   (a) *Sort the criticalities for all requirements.*
   (b) *Arrange requirements from top down in a descending order of the criticalities. Requirements with the same criticality will be placed at the same level.*
2. *Top-level requirements:*
   (a) *If there is only one requirement with the highest criticality, place it on the top of the hierarchy.*
   (b) *Else if there are more than one requirement with the highest criticality,*

      i. Compute the total cooperative degree for each requirement with the rest of the requirements whose criticality is the same.

     ii. Sort the total cooperative degrees computed in the previous step.

     iii. Arrange requirements from left to right on the top in a descending order of the total cooperative degrees.

     iv. Add a virtual requirement on the top of those requirements.

3. *Grouping requirements (between two adjacent levels):*
   (a) *For requirements (other than requirements at the top level) with the same criticality, compute all the cooperative degrees for each requirement at level i with every requirement at level $i - 1$.*
   (b) *Given a requirement, R_h, at level i, for every requirements, R_k, at level $i - 1$, sort the cooperative degree of R_h and R_k, obtained from the previous step. Group the requirement whose cooperative degree is the highest under R_h.*
   (c) *Continue the previous step until all the requirements at level i have been considered.*

4. *Left-right (for each group):*
   (a) *Given a requirement, R_k, at level $i - 1$, for every requirements, R_h, at level i, sort the cooperative degrees of R_h and R_k.*
   (b) *Arrange from left to right the requirements at level i in a descending order.*

This step is important in the sense that the ordering established through the hierarchy helps to alleviate the associativity problem inherited in averaging operators, namely, a unique ordered list can thus be obtained.

Select aggregation operators. By applying Theorem 1, given requirement $R = <\varphi_1, \varphi_2>$, $\mu_R(x = u_i) = fhold(\varphi_2, a)$, $a \in A_b$, we use μ_R instead of $\mu_R(x = u_i)$, where $x \in X$. In the following a reasonable set of properties for the aggregation operations is proposed in order to characterize the aggregation rule :

1. Idempotency: given two requirements R_1, R_2, aggregation operation ϕ : $[0, 1]^2 \rightarrow [0, 1]$; if the both are same requirement, that is $R_1 = R_2$, $\forall\ a \in A_b$, $\mu_{R_1} = \mu_{R_2} = \alpha$; then

$$\phi(\mu_{R_1}, \mu_{R_2}) = \alpha$$

2. Intended relationship: given any permutation of the set of requirements $\pi: 1,...,n \rightarrow 1,...,n$ if $r \in [cf, cp, ir, cb]$, $r(R_1, R_2) \neq (R_2, R_3)$ then

$$\phi(\mu_{R_1}, \mu_{R_2}, \mu_{R_3}) \neq \phi(\mu_{R_{\pi_1}}, \mu_{R_{\pi_2}}, \mu_{R_{\pi_3}})$$

3. Satisfiability: if $\mu_{R_1} \geq \mu_{R_2}$, $\mu_{R_3} \geq \mu_{R_4}$ and $r(R_1, R_3) = r(R_2, R_4)$ then

$$\phi(\mu_{R_1}, \mu_{R_2}) \geq \phi(\mu_{R_2}, \mu_{R_4})$$

4. Criticality: the criticality of R_1 is c_1, the criticality of R_2 is c_2, then

$$\phi(\mu_{R_1(c_1)}, \mu_{R_2(c_2)}) = \phi(c_1 \otimes \mu_{R_1}, c_2 \otimes \mu_{R_2})\ where \otimes\ is\ a\ fuzzy\ multi\text{-}$$
$$plication\ operator.$$

To select appropriate aggregation operators, we propose the consideration of operators that can: (1) reflect the intended relationship between requirements, (2) associate the criticality with each requirement, (3) fit the aggregation structure, and (4) incorporate the notion of conflicting and cooperative degrees. We have chosen *fuzzy and* and *fuzzy or* operators proposed in [44] due to the fact that *fuzzy and* operator can be used within each conjunction, while *fuzzy or* can be applied between each disjunction, and that the *compensation* between aggregated sets can be achieved by incorporating the conflicting and cooperative degrees into the parameters in these two operators, which in turn can reflect the relationships between requirements. In requirements trade-off analysis approach, all the requirements are considered during the aggregation process, and hence no requirement will be excluded out. However, the lower the degree of the criticality of a requirement, the lower the impact of the requirement on the result of the aggregation. Therefore, the exclusive or aggregation is not considered in requirements trade-off analysis approach. In addition, the mathematical structure of these operators is easy and can be handled efficiently [44]. In order to better match our aggregation structure, these two operators are also adopted for criticalities aggregation to reflect different relationships between the requirements. We formally define these two operators below.

Definition 6. *(Extended Fuzzy and)*
Assume two requirements $< w_{R_1}, \mu_A(x), R_1 >, < w_{R_2}, \mu_B(x), R_2 >, \forall x \in X$.

$$< w_{R_1}, \mu_A(x), R_1 > \wedge_{\gamma_{and}} < w_{R_2}, \mu_B(x), R_2 > \quad \text{is defined as}$$

$$< w_{and}(w_{R_1}, w_{R_2}), \mu_{and}(\mu_A(x), \mu_B(x)), R_1 \wedge R_2 >$$

where

$$\mu_{and}(\mu_A(x), \mu_B(x)) = \gamma_{and} \ min\{\mu_A(x), \mu_B(x)\} + \frac{(1-\gamma_{and})(\mu_A(x)+\mu_B(x))}{2}$$

$$w_{and}(w_{R_1}, w_{R_2}) = \gamma_{and} \ min\{w_{R_1}, w_{R_2}\} + \frac{(1-\gamma_{and})(w_{R_1}+w_{R_2})}{2}$$

$$\gamma_{and} = (cf - cp + 1)/2 \ and \ \gamma_{and} \in [0,1].$$

Definition 7. *(Extended Fuzzy or)*
Assume two requirements $< w_{R_1}, \mu_A(x), R_1 >, < w_{R_2}, \mu_B(x), R_2 >, \forall x \in X$.

$$< w_{R_1}, \mu_A(x), R_1 > \vee_{\gamma_{or}} < w_{R_2}, \mu_B(x), R_2 > \quad \text{is defined is}$$

$$< w_{or}(w_{R_1}, w_{R_2}), \mu_{or}(\mu_A(x), \mu_B(x)), R_1 \vee R_2 >$$

where

$$\mu_{or}(\mu_A(x), \mu_B(x)) = \gamma_{or} \ max\{\mu_A(x), \mu_B(x)\} + \frac{(1-\gamma_{or})(\mu_A(x)+\mu_B(x))}{2}$$

Table 3.2. Relationships vs Aggregation Operators

relation / operator	conflicting			irrelevant	counterbalance	cooperative		
	strong	moderate	weak			strong	moderate	weak
$\wedge$ $(\gamma_{and}=(cf-cp+1)/2)$	min	fuzzy-and		fuzzy-and		arith mean	fuzzy-and	
	$\gamma_{and}=1$	$1>\gamma_{and}>0.5$	$1>\gamma_{and}>0.5$	$(\gamma_{and}=0.5)$		$\gamma_{and}=0$	$0.5>\gamma_{and}>0$	$0.5>\gamma_{and}>0$
$\vee$ $(\gamma_{or}=(cp-cf+1)/2)$	arith mean	fuzzy-or		fuzzy-or		max	fuzzy-or	
	$\gamma_{or}=0$	$0.5>\gamma_{or}>0$	$0.5>\gamma_{or}>0$	$(\gamma_{or}=0.5)$		$\gamma_{or}=1$	$1>\gamma_{or}>0.5$	$1>\gamma_{or}>0.5$

$$w_{or}(w_{R_1}, w_{R_2}) = \gamma_{or} \ max\{w_{R_1}, w_{R_2}\} + \frac{(1-\gamma_{or})(w_{R_1}+w_{R_2})}{2}$$

$$\gamma_{or} = (cp - cf + 1)/2 \ and \ \gamma_{or} \in [0, 1].$$

In the case that two requirements are connected by *and*, there are two situations: (1) if γ_{and} equals to 1 (i.e. requirements are strongly conflicting), the *fuzzy and* operator reduces to *min*, and (2) if γ_{and} equals to 0 (i.e. requirements are strongly cooperative), the operator becomes *arithmetic mean*. If the requirements are connected by *or*, the *fuzzy or* operator yields *max* under the condition that γ_{or} equals to 1; whereas, the operator boils down to *arithmetic mean* if γ_{or} equals to 0. Although the ranges for the parameters γ_{and}/γ_{or} in the case of "moderate" and "weak" relationships are similar, a subtle distinction can still be made through the observation of *cf* and *cp*. That is, in the case of "moderate" relationship, $cf + cp = 1$; whereas, for "weak" relationship, $cf + cp < 1$. This distinction can have some impact on the computation of γ_{and} and γ_{or}.

$\gamma_{and} = \gamma_{or} = 0.5$ indicates that two requirements are either *irrelevant* or *counterbalance*. Generally, if two requirements are *independent*, there is no need for performing the aggregation. These are summarized in table 3.2.

The *fuzzy or* and *fuzzy and* operator can be easily proved to satisfy aggregation properties proposed. :

1. Idempotency: given two requirements R_1, R_2, $\mu_{R_1} = \mu_{R_2} = \alpha$; then

$$\mu_{and}(\mu_{R_1}, \mu_{R_2}) = (\mu_{R_1} + \mu_{R_2})/2 = \alpha \ , \gamma_{and} = 0$$

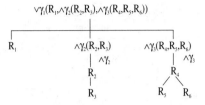

Fig. 3.3. An Example of the Extended Hierarchical Aggregation Structure

$$\mu_{or}(\mu_{R_1}, \mu_{R_2}) = max(\mu_{R_1} + \mu_{R_2}) = \alpha \, \gamma_{or} = 1$$

2. Intended relationship: given three requirements R_1, R_2, and R_3, $r(R_1, R_2) \neq (R_2, R_3)$ then

  assume $\gamma_{and}(R_1, R_2) = \alpha$, $\gamma_{and}(R_2, R_3) = \beta$;

$$\mu_{and}(\mu_{and}(\mu_{R_1}, \mu_{R_2}), \mu_{R_3}) \neq \mu_{and}(\mu_{R_1}, \mu_{and}(\mu_{R_2}, \mu_R 3))$$

  assume $\gamma_{or}(R_1, R_2) = \alpha$, $\gamma_{or}(R_2, R_3) = \beta$;

$$\mu_{or}(\mu_{or}(\mu_{R_1}, \mu_{R_2}), \mu_{R_3}) \neq \mu_{or}(\mu_{R_1}, \mu_{or}(\mu_{R_2}, \mu_R 3))$$

3. Satisfiability: if $\mu_{R_1} \geq \mu_{R_2}$, $\mu_{R_3} \geq \mu_{R_4}$ and $r(R_1, R_3) = r(R_2, R_4)$ then

$$\gamma_{and}(R_1, R_3) = \gamma_{and}(R_2, R_4), \; \mu_{and}(\mu_{R_1}, \mu_{R_3}) \geq \mu_{and}(\mu_{R_2}, \mu_{R_4})$$

$$\gamma_{or}(R_1, R_3) = \gamma_{or}(R_2, R_4), \; \mu_{or}(\mu_{R_1}, \mu_{R_3}) \geq \mu_{or}(\mu_{R_2}, \mu_{R_4})$$

4. Criticality: the criticality of R_1 is c_1, the criticality of R_2 is c_2, then

$$\mu_{and}(\mu_{R_1}, \mu_{R_2}) = \mu((c1/c1)\, \mu_{R_1}, (c_2/c1)\, \mu_{R_2})$$

$$\mu_{or}(\mu_{R_1}, \mu_{R_2}) = \mu((c1/c1)\, \mu_{R_1}, (c_2/c1)\, \mu_{R_2})$$

Aggregate requirements. To aggregate requirements in a requirement hierarchy, there are two steps involved: (1) to utilize the breadth first search algorithm to traverse the requirements hierarchy to form an ordered list, and (2) to apply *fuzzy and* or *fuzzy or* operator recursively to the requirements in the list.

Definition 8. *(Hierarchy aggregation rule [10])*:
Let $\phi_0, \phi_1, ..., \phi_c$ be c+1 aggregation operators, ϕ_0 has dimension c; ϕ_i has

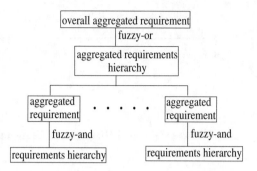

Fig. 3.4. The Extended Hierarchical Aggregation Structure

dimension h_i, $h_i \geq 1$, i=1,...,c, such that $\sum_{i=1}^{c} h_i = n$ and $\phi_i : [0,1]^{h_i} \to [0,1]$, in such a way that the composition

$$\phi \equiv \phi_0(\phi_1, ..., \phi_c) : [0,1]^c \to [0,1]$$

defined as follows

$$\phi_0(\phi_1, ..., \phi_c)(\mu_{R_1}, ..., \mu_{R_n}) = \phi_0(\phi_1(\mu_{R[1]}, ..., \mu_{R[n]}), ..., \phi_c(\mu_{R[n-h_c+1]}, ..., \mu_{R[n]}))$$

$$\forall R_i, \; i = 1..n$$

is an hierarchy aggregation rule.

Finally, a four-level hierarchical aggregation structure can thus be built (see fig. 3.4):

1. Requirements hierarchies built from disjuncts by applying Algorithm 1 are placed at the bottom of the hierarchical structure. Each requirement hierarchy is converted into its ordered list.
2. *Fuzzy and* operator is applied recursively to glue requirements in an ordered list to establish an aggregated requirement, which is placed on the top of the requirements hierarchy.
3. All the aggregated requirements will be used to build an aggregated requirements hierarchy, in which the aggregated requirement in the hierarchy are in turn combined using *fuzzy or* operator recursively to form an overall aggregated requirement. It is of interest to note that a hierarchy in requirements trade-off analysis approach is established mainly based on the relationships among the requirements and the criticalities of the requirements. Therefore, only the change of relationships and criticalities will result in the change of the hierarchy. Namely, given a new set of relationships and criticalities, requirements trade-off analysis approach will re-build a new hierarchy instead of reusing the previous hierarchy.

3.5   Related Work

Work in a number of fileds has made its mark on our requirements trade-off analysis framework. Analogies of trade-off analysis based on relationships may be found in multi-criteria decision making [7,18], AI planning [31,42], and multi-perspectives specifications [48,16,14,20,34].

Multi-criteria Decision Making. Relationship analysis approaches in multi-criteria decision making [6,7,19,17] focus on an explicit modeling of relationships between goals, and on the determination of the final set of decision alternatives according to the relationships. Carlsson and Fuller [7] propose an approach to fuzzy multiple objective programming (FMOP) with interdependency relationships among objectives, which is an extension of Carlsson's MOP [6] to fuzzy logic. Three kinds of relationships have been identified: supportive, conflicting, and independent. The basic idea is to utilize these relationships to modify the membership function of the so called "good solution". Felix *et al.* [17,19] propose an approach, called DMRG (Decision Making Based on Relationship between Goals), to defining a spectrum of relationships based on fuzzy inclusion and fuzzy non-inclusion: independent, assist, cooperate, analogous, hinder, compete, trade-off and unspecified dependent, and to determining the final set of decision alternatives according to the relationships. These approaches are similar to ours in two aspects: the problems of modeling the relationships, and the issues of aggregation.

AI Planning. Research in the areas of goals conflict in AI planning tackles issues similar to the trade-off analysis, for example,

- Sycara [42] provides a negotiation method, called PERSUADER, to find a compromise acceptable to all agents under the situations that planning goals are ill-specified, subgoals cannot be enumerated and the utilities with the subgoals are not precisely known. The negotiation is performed through proposal and modification of goal relaxation. There are two ways of reacting to negative feedback through negotiation: changing the rejecting agent's evaluation of the plan through persuasive argumentation, and modifying / repairing the plan so that it will be more acceptable. The main difference between PERSUADER and our approach in the trade-off analysis is that in aggregating individual requirements, requirements are compensated to each other based on their relationships in our approach; whereas, PERSUADER needs to modify user's utility if no solution can be obtained.

- Luria [31] proposes a commonsense planner called KIP (Knowledge Intensive Planner) which is developed for the Unix Consultant system. KIP uses goals conflict concerns to deal with potential goal conflicts. Luria classifies goal conflict concerns into six types: default concerns, violated-default concerns, intended effects concerns, unintended effects concerns, expressed goal conflict concerns and effect goal conflict concerns. After KIP detects the goals of the user, it selects a potential plan. KIP then checks for violated defaults goal conflict concern. KIP next proceeds the intended effect of the selected plan about user goal. Finally, KIP evaluates the degree of those concerns. If the degree of concern is low, KIP disregards the concern. If the degree of concern is high, KIP elevates the concern to a source of plan failure and pass it to goal conflict resolution. Conflict resolution may occur by either modifying the plan, or choosing a new plan.

Multi-Perspectives Specifications. Work on multiple perspectives has been investigated along several directions. Feather [16] suggests using many parallel evolutionary transformations for constructing specifications, which may then be merged by replaying them sequently. Finkelstein and Fuks [20] develop a framework to support the construction of formal specifications and reasoning about the process of specifications from multiple viewpoints. Their model has two parts: an underlying viewpoint architecture and a dialogue scheme, which combines the dialogue logics with cooperation and negotiation approaches. Dialogues are used to perform viewpoints negotiation, to establish responsibilities of participants, and to construct an overall specification in a cooperative manner among the participants. The viewpoint architecture includes viewpoint, commitment store, working area, event store and dialogue kernel. A viewpoint is a logical participant in the dialogue. A physical participant in a dialogue may present many logical viewpoints. Each viewpoint has a commitment store which holds it's commitments within the dialogue. The dialogue scheme is presented in terms of three constructs: acts, events and commitments.

Easterbrook [14] proposes a framework for representing conflicting viewpoints in a domain model. A viewpoint in his framework is a self-consistent description of an area of knowledge representing the context in which a role is performed. In evolving viewpoints, a new viewpoint will need to be split if it causes inconsistency. The new viewpoint and its negation are placed in different descendants of the original viewpoint, so that each remains self-consistent individually. The detection of conflict might be based on detection of logical inconsistencies. Thus, a hierarchy of viewpoints is established as the elicitation proceeds. The inheritance structure implies that the higher an item in the hierarchy, the more widely agreed it is. One of the aims of using viewpoints is to reduce the need for consistency checks. This approach allows many viewpoints to be combined into a single domain model without necessary resolving conflicts between them. Conflicts are treated as an important part of the domain, and are to be represented and understood.

Robinson and Fickas [37,38] propose an approach, called Oz, to requirements negotiation. There are three steps involved in Oz: conflict detection, resolution generation and resolution selection. The conflict detector of Oz does a pairwise comparison across all specifications. It does so by matching up design goals from perspectives and by comparing their plans. The specifications and conflicts will be passed to conflict resolver which will provide analytic compromise and heuristic compensation and dissolution for each conflict. Compensation is to add similar but conflict free requirements to negotiations, while, dissolution is to replace conflicting items potentially less contentious items. Finally, the resolver will provide guidance for search control by choosing intermediate alternatives and automated negotiation methods. Each method can be applied in any sequence to derive resolutions. The nonconflicting specifications are jointed into a single specification by merger of Oz.

Furthermore, our notion of conflicting and cooperative degrees can be interpreted as taking the distance-wise view between any two individual requirements; whereas, Yen et al. [48] addressed the same issue from the probability aspects of conflicting and cooperative relationships, that is, computing the ratio between the total number of after state pairs and those of conflicting or cooperative pairs. No irrelevant pair is considered.

3.6   Conclusion

As was pointed by Lamsweerde [43], goal information should be captured in the requirements acquisition phase, which is useful for analyzing conflicting requirements and nonfunctional requirements. The proposed approach is spawned based on this belief to fuse the goal-oriented and object-oriented modeling techniques in requirements engineering.

Our approach offers several benefits: (1) serving as a structuring mechanism to facilitate the derivation of use case specifications and objects model; (2) bridging the gap between the domain description and the system requirements, that is, the interactions between functional and nonfunctional requirements; (3) making easy the handling of soft requirements, and the analysis of conflicting requirements; and (4) extending traditional object-oriented techniques to fuzzy logic to manage different kinds of fuzziness that are rooted in user requirements.

Our future research plan will consider the following task: to investigate the worth-oriented domain advocated by Rosenschein et a. [39] to evaluate dynamic behaviors without actually executing the statechart diagrams.

References

1. A.I. Anton. Goal-based requirements analysis. In *Proceedings of the International Conference on Requirements Engineering*, pages 136–144, 1996.
2. R. Balzer, N. Goldman, and D. Wile. Informality in program specifications. *IEEE Transactions on Software Engineering*, 4(2):94–103, 1978.
3. M. Barbacci, T.H. Longstaff, M.H. Klein, and C.B. Weinstock. Quality attributes. Technical Report CMU/SEI-95-TR-021, CMU, Dec. 1995.
4. B.I. Blum. Representing open requirements with a fragment-based specification. *IEEE Transactions on Software Engineering*, 23(3):724–736, 1993.
5. A. Borgida, S. Greenspan, and J. Mylopoulos. Knowledge representation as the basis for requirements specification. *Computer*, pages 82–91, April 1985.
6. C. Carlsson. On optimization with interpendent multiple criteria. In R. Lowen and M. Roubens, editors, *Fuzzy Logic: State of the Art*, pages 287–300. Kluwer Academic Publishers, Netherlands, 1993.

7. C. Carlsson and R. Fuller. Interdependence in fuzzy multiple objective programming. *Fuzzy Sets and Systems*, 65:19–29, 1994.
8. A. Cockburn. Goals and use cases. *Journal of Object-Oriented Programming*, 10(7):35–40, September, 1997.
9. V. Cutello and J. Montero. A characterization of rational amalgamation operations. *International Journal of Approximate Reasoning*, 8:325–344, 1993.
10. V. Cutello and J. Montero. Hierarchies of intensity preference aggregations. *International Journal of Approximate Reasoning*, 10:123–133, 1994.
11. V. Cutello and J. Montero. The associativity problem for owa operators. In *Proceedings of 6th International Fuzzy Systems Association World Congress*, pages (I)149–152, 1995.
12. A. Dardenne, A. van Lamsweerde, and S. Fickas. Goal-directed requirements acquisition. *Science of Computer Programming*, 20:3–50, 1993.
13. A.M. Davis, editor. *Software Requirements: Analysis & Design*. Englewood Cliffs, NJ: Prentice-Hall, 1990.
14. S. Easterbrook. Domain modelling with hierarchies of alternative viewpoints. In *Proceedings of the IEEE International Symposium on Requirements Engineering*, pages 65–72, 1993.
15. R.O. Flamm et al. The integrated southern pine beetle expert systems. *Expert Systems with Applications*, 2:97–105, 1991.
16. M.S. Feather. Constructing specifications by combining parallel elaboration. *IEEE Transactions on Software Engineering*, 15(2):198–208, 1989.
17. R. Felix. Relationships between goals in multiple attribute decision making. *Fuzzy Sets and Systems*, 67:47–52, 1994.
18. R. Felix. Fuzzy decision making based on relationships between goals compared with the analytic hierarchy process. In *Proceedings of 6th International Fuzzy Systems Association World Congress*, pages 253–256, 1995.
19. R. Felix, S. Reddig, and A. Adelhof. Multiple attribute decision making based on fuzzy relationships between objectives and its application in metal forming. In *Proceedings of the Second IEEE International Conference on Fuzzy Systems*, pages 378–383, 1993.
20. A. Finkelstein and H. Fuks. Multi-party specification. In *Proceedings of the International Workshop on Software Specifications and Design*, pages 185–195, 1989.
21. W.L. Johnson. Knowledge-based software engineering. In A. Kent and J.G. Williams, editors, *Encyclopedia of Computer Science and Technology*, volume 31, pages 173–225. Marcel Dekker, New York, 1994.
22. U. Kaymak and H.R. van Nauta Lemke. Selecting an aggregation operator for fuzzy decision making. In *Proceedings of Third IEEE International Conference on Fuzzy Systems*, pages 1418–1422. IEEE Press, 1994.
23. G. Kotonya and I. Sommerville. Viewpoints for requirements definition. *Software Engineering Journal*, pages 375–387, Nov. 1992.
24. J. Lee and J.Y. Kuo. New approach to requirements trade-off analysis for complex systems. *IEEE Transactions on Knowledge and Data Engineering*, 10(4):551–562, July/August 1998.
25. J. Lee, J.Y. Kuo, and W.T. Huang. Classifying, analyzing and representing informal requirements. In *Proceedings of the Sixth International Fuzzy Systems Association World Congress (IFSA'95)*, pages (I)645–648, July 1995.
26. J. Lee, L.F. Lai, and W.T. Huang. Task-based specifications through conceptual graphs. *IEEE Expert*, 11(4):60–70, Aug. 1996.

27. J. Lee and N.L. Xue. Analyzing user requirements by use cases: A goal-driven approach. *IEEE Software*, 16(4):92–101, July/August 1999.
28. J. Lee, N.L. Xue, and J.Y. Kuo. Structuring requirement specifications with goals. *Information and Software Technology*, pages 121–135, February 2001.
29. P. Loucopoulos and V. Karakostas. *System Requirements Engineering*. McGraw-Hill, Berkshire, UK, 1995.
30. M.R. Lowry and R.D. McCartney, editors. *Automating Software Design*. AAAI Press, Menlo Park, CA, 1991.
31. M. Luria. Goal conflict concerns. In *Proceedings of the 12th International Joint Conference on Artificial Intelligence*, pages 1025–1031, 1987.
32. J. McCarthy and P. Hayes. Some philosophical problems from the standpoint of artificial intelligence. In *Machine Intelligence 4*, pages 463–502. Edinburgh University Press, Scotland, 1969.
33. Celso Niskier, Tom Maiibaum, and Daniel Schwabe. A look through prisma: Towards pluralistic knowledge-based environments for software specification acquisition. In *5th International Workshop on Software Specification and Design*, pages 128–136, May, 1989.
34. B. Nuseibeh, J. Kramer, and A. Finkelstein. A framework for expressing the relationships between multiple views in requirements specification. *IEEE Transactions on Software Engineering*, 20(10):760–773, Oct. 1994.
35. C. Potts, K. Takahashi, and A.I. Anton. Inquiry-based requirements analysis. *IEEE Software*, 11(2):21–32, March 1994.
36. H.B. Reubenstein and R.C. Waters. The requirements apprentice: Automated assistance for requirements acquisition. *IEEE Transactions on Software Engineering*, 17(3):226–240, 1991.
37. W.N. Robinson. Negotiation behavior during requirement specification. In *Proceedings of 12th International Conference on Software Engineering*, pages 268–276, 1990.
38. W.N. Robinson and S. Fickas. Supporting multi-perspective requirements engineering. In *Proceedings of first International Conference on requirement engineering*, pages 206–215. IEEE Computer Society, 1994.
39. J.S. Rosenschein and G. Zlotkin. *Rules of Encounter*. The MIT Press, Cambridge, Massachusetts, 1994.
40. T. L. Saaty. *Decision Making for Leaders: The Analytic Hierarchy Process for Decisions in a Complex World*. Lifetime Learning, Atlanta, Georgia, 1982.
41. P.A. Subrahmanyam. The software engineering of expert systems. *IEEE Transactions on Software Engineering*, 11(11):1391–1400, Nov. 1985.
42. K. Sycara. Resolving goal conflicts via negotiation. In *Proceedings of the 6th National Conference on Artificial Intelligence*, pages 245–250. Cambridge, Mass., MIT Press, 1988.
43. A. van Lamsweerde, R. Darimont, and P. Massonet. Goal-directed elaboration of requirements for a meeting scheduler problems and lessons learnt. Technical Report RR-94-10, Universite Catholique de Louvain, Departement d'Informatique, B-1348 Louvain-la-Neuve, Belgium, 1994.
44. B.M. Werner. Aggregation models in mathematical programming. In G. Mitra, editor, *Mathematical Models for Decision Support*, pages 295–305. Springer-Verlag, Berlin, 1988.
45. R.R. Yager and D.P. Filev. On the extension of owa operators. In *Proceedings of the Sixth International Fuzzy Systems Association World Congress (IFSA'95)*, pages (II)161–163, 1995.

46. J. Yen and J. Lee. Fuzzy logic as a basis for specifying imprecise requirements. In *Proceedings of IEEE International Conference on Fuzzy Systems*, pages 745–749, 1993.

47. J. Yen and J. Lee. A task-based methodology for specifying expert systems. *IEEE Expert*, 8(1):8–15, Feb. 1993.

48. J. Yen, X. Liu, and S.H. Teh. A fuzzy logic-based methodology for the acquisition and analysis of imprecise requirements. *Concurrent Engineering: Research and Applications*, 2:265–277, 1994.

49. L.A. Zadeh. Fuzzy set as a basis for a theory of possibility. *Fuzzy Sets and Systems*, 1:3–28, 1978.

50. L.A. Zadeh. Test-score semantics as a basis for a computational approach to the representation of meaning. *Literacy Linguistic Computing*, 1:24–35, 1986.

51. H.-J. Zimmermann. *Fuzzy Set Theory and Its Applications*. Kluwer Academic, Boston, MA, 1991.

52. H.-J. Zimmermann and P. Zysno. Decisions and evaluations by hierarchical aggregation of information. *Fuzzy Sets and Systems*, 10:243–260, 1983.

4 A Generalized Object-Oriented Data Model Based on Level-2 Fuzzy Sets

Guy de Tré, Rita de Caluwe, Jörg Verstraete, and Axel Hallez

Ghent University, Dept. of Telecommunications and Information Processing,
Sint-Pietersnieuwstraat 41, B-9000 Gent, Belgium

4.1 Introduction

Classical software systems can almost exclusively deal with crisp information. However, in the real-world, a large amount of all the information is merely available in a fuzzy and/or uncertain form. Therefore, information modeling and handling in classical software applications is subject to a loss of semantics. This problem is especially recognized in the database and knowledge base research communities and is seen as a challenge for the near future [15].

In order to tackle the loss of data semantics, future software engineering tools need to be able to cope with information in its most general form, i.e. with both "fuzzy" and "crisp" information.

"Fuzzy" information can be classified as follows [17,18]:

- *Uncertain information*, i.e. information for which it is not possible to determine whether it is true or false. The concept of uncertainty denotes a lack of knowledge about the *truth value* of the available information.
- *Imprecise information*, i.e. information which is not available as specific as it should be. The concept of imprecision denotes a lack of knowledge about the *accuracy* of the available information, which is *explicitly* specified.
- *Vague information*, i.e. information which is inherently vague. The concept of vagueness denotes a lack of knowledge about the *accuracy* of the available information, which is *implicitly* specified.
- *Inconsistent information*, i.e. information which contains two or more assertions that can not be true at the same time.
- *Incomplete information*, i.e. information for which some data are missing.

The concepts of *uncertainty* and *imprecision* both cover the notion of a deficiency of information due to a lack of knowledge. Although there exists a certain affinity among these concepts, they are orthogonal: they can appear side by side or simultaneously. For example, the age of a person can be imprecise and be given as *"between 15 and 20 years old"*. The statement that a person is *"possibly 20 years old"* is an example of uncertain information, whereas uncertainty can also occur in combination with imprecision as in e.g. the age of the person is *"possibly between 15 and 20 years old"*.

The concept of *vagueness* is associated with the inability to define sharp or precise borders for some domain (of information) and therefore represents an inherent uncertainty. Examples of vague descriptions of, e.g. the age of a person are the concepts *"young"*, *"middle-aged"* and *"old"*. These concepts are multi-valued concepts as, e.g., a number of different ages can be considered as *"young ages"*. Moreover, these concepts are vague concepts because of the borderline between whether or not a value belongs to the values that are described by the concept, is not uniquely nor well defined. Vague concepts can be used to express both imprecision (e.g. the person is *"young"*) and the combination of both uncertainty and imprecision (e.g. the person in possibly *"young"*).

Inconsistency describes a situation in which two or more items come into conflict, e.g. *"Joe is 16 years old"* and *"Joe is older than 20 years"*. In such a situation there is no possibility to find a consensus for the conflicting items to coexist. The only solution is to omit the information of the less reliable source (assumed that this source is known).

A lack of information is indicated with the concept of *incompleteness*. Incompleteness can result from the unavailability and/or the irrelevancy of the information —in contrast with uncertainty and imprecision where information is (partly) available—.

"Crisp" information can be seen as a special kind of "fuzzy" information in which no fuzziness is involved [6]. In fact, one has to realize that the collection of all "crisp" information is only a small part of the collection of information available in the "real-world" (see Fig. 4.1).

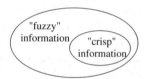

Fig. 4.1. Crisp information as a special kind of ("real-world") information.

Until several years ago, software development has been mainly concentrated onto the representation and handling of "crisp" data and the underlying idea that (complex) "real-world" data can be approximately represented as crisply described data. This implies that a lot of software applications can still be improved by adapting more advanced modeling techniques, which allow for a more adequate data representation and data manipulation.

Nowadays, the mathematical foundations for the definition of such advanced modeling techniques exist. A turning-point was reached with the introduction of multi-valued logics [2,19] which, on their turn, have been the onset for the fuzzy set theory [21], possibility theory [24,7] and fuzzy logic [23]. These theories have been well developed and are very appropriate for the formal handling of fuzzy and uncertain information. More recently, the introduction of "computing with words" [25], of the "theory of fuzzy information granulation" [26] and of the "computational theory of perceptions" [27] incites the development of new general information modeling techniques.

The object-oriented modeling approach presented in this chapter is based on fuzzy set theory (more specifically on the concept of a level-2 fuzzy set) and on the theories of "computing with words" and of "fuzzy information granulation" (more specifically on the concept of a generalized constraint). The presented data model is a generalization of an ordinary data model, which distinguishes it from other approaches presented in the literature [9,1,4,3] that rather involve the extension of an ordinary data model. One of the main advantages of the presented "generalization"-approach is that it allows for a uniform representation and manipulation of data in its most general form (i.e. crisp, uncertain, imprecise, vague, incomplete, and any valid combination of these).

This chapter is organized in four sections. Section 4.1 is the introductory section. In section 4.2, a full semantic definition of the concept "type" is given. A type is a generic concept, which informally can be seen as a description of the common characteristics of a data collection and is generally acknowledged to be a suitable means for the modeling and the handling of crisp information. Furthermore, an object-oriented "type system" is presented. This type system provides a formal framework for the definition of so-called object types, which in accordance with the object-oriented paradigm [14], are acknowledged to be a suitable means for the modeling of both the structure and the behavior of "crisp" information. The presented type system is consistent with the OMG-object model [12,20] specifications, which guarantees its compatibility with all "OMG-compliant" object-oriented programming languages. The semantic definition of the concept "type" is generalized in the first part of section 4.3. This is done by using level-2 fuzzy sets and results in the definition of the new concept "generalized type". Generalized types provide a means to handle uncertain, imprecise and vague information as well. The generalization of the object-oriented type system is described in the second part of section 4.3. It is shown how generalized constraints, as defined by L.A. Zadeh [25,26], provide for a suitable means to define the semantics of the components of generalized (object) types. Finally, a summary of the obtained model and future developments are presented in the concluding section.

4.2   Types and Type System

Computer software is useful because it models "real-world" processes. To do this, a computer program manipulates abstract entities that represent "real-world" entities. The closer the characteristics of an abstract entity mirror those of the corresponding "real-world" entity, the more effective will be the program. The common characteristics of a collection of abstract entities can be described by means of a type. This is the main reason why most data models, including the data model presented in this chapter, support some "type"-notion. The concept of a type is formally defined in subsection 4.2.1.

In order to define the types supported by a given data model, a type system can be used [16]. A type system can informally be seen as a collection of rules, which together define the supported types. The type system for the presented data model is formally defined in subsection 4.2.2.

4.2.1   Types

Each type is defined by a type specification and one or more type implementations. In order to give a complete definition of the type specification, it is necessary to provide the rules which define its syntax, as well as the rules which define its semantics.

- *The syntax of a type (specification):* The syntax rules for a type specification can be formally described by means of some mathematical expressions.
- *The semantics of a type (specification):* It has been stated in [10] that the semantic definition of a type with specification *ts* can be fully determined by:
  - a set of domains D_{ts},
  - a designated domain $dom_{ts} \in D_{ts}$,
  - a set of operators O_{ts},
  - and a set of axioms A_{ts}.

  The designated domain dom_{ts} defines the set of the valid values for the type —also called the instances of the type— and is called the domain of the type. In order to deal with "undefined" domain values in a formal way, the assumption has been made that every domain dom_{ts} contains a special, domain specific value $\perp_{ts}$, which is used to represent "undefined" domain values. The set of operators O_{ts} contains the operators, which are implicitly defined on the domain dom_{ts}. The set of domains D_{ts} consists of the domains that are involved in the definition of the operators of O_{ts}, whereas the set of axioms A_{ts} contains the axioms, which are involved in the definition of the semantics of the operators of O_{ts}.

Example 1. The type specification, which describes the common characteristics of the integer values is fully determined as follows:

- *Syntax.* The type for the modeling of integer values is represented by the symbol (keyword) *Integer*.

- *Semantics.* The semantics of the *Integer* type have been defined by:
  - The designated domain $dom_{Integer}$:

  $$dom_{Integer} = \mathbb{Z} \cup \{\perp_{Integer}\}$$

  where $\mathbb{Z}$ represents the set of the integer values.

– The set of domains $D_{Integer}$:

$$D_{Integer} = \{dom_{Boolean}, dom_{Integer}\}$$

where $dom_{Boolean}$ is the domain of the type, which describes the characteristics of the Boolean values (i.e. *True* and *False*). The domain $dom_{Boolean}$ is necessary for the definition of the comparison operators.

– The set of operators $O_{Integer}$:

$$\begin{aligned}
O_{Integer} = \{ &= : (dom_{Integer})^2 \rightarrow dom_{Boolean}, \\
&\neq : (dom_{Integer})^2 \rightarrow dom_{Boolean}, \\
&< : (dom_{Integer})^2 \rightarrow dom_{Boolean}, \\
&> : (dom_{Integer})^2 \rightarrow dom_{Boolean}, \\
&\leq : (dom_{Integer})^2 \rightarrow dom_{Boolean}, \\
&\geq : (dom_{Integer})^2 \rightarrow dom_{Boolean}, \\
&+ : (dom_{Integer})^2 \rightarrow dom_{Integer}, \\
&- : (dom_{Integer})^2 \rightarrow dom_{Integer}, \\
&* : (dom_{Integer})^2 \rightarrow dom_{Integer}, \\
\text{div} &: (dom_{Integer})^2 \rightarrow dom_{Integer}, \\
\text{mod} &: (dom_{Integer})^2 \rightarrow dom_{Integer}, \\
&\bot : \rightarrow dom_{Integer}\}
\end{aligned}$$

– The set of axioms $A_{Integer}$:

* The semantics of the binary operators $=$, $\neq$, $<$, $>$, $\leq$ and $\geq$: as $dom_{Integer} \setminus \{\perp_{Integer}\}$ equals to the set $\mathbb{Z}$ —which is totally ordered—, it is stated that the binary operators $=$, $\neq$, $<$, $>$, $\leq$ and $\geq$ have the same semantics as their counterparts $=$, $\neq$, $<$, $>$, $\leq$ and $\geq$ within $\mathbb{Z} \times \mathbb{Z}$, when restricted to the set:

$$dom_{Integer} \setminus \{\perp_{Integer}\}$$

For undefined argument values the semantics become:

$$\forall\ x \in dom_{Integer}:$$
$$relat\ (x, \perp_{Integer}) = relat\ (\perp_{Integer}, x) = \perp_{Boolean}$$

where $relat$ is a variable copula whose values are respectively the symbols $=$, $\neq$, $<$, $>$, $\leq$ and $\geq$.

* The semantics of the binary operators $+$, $-$, $*$, div and mod: for the same reason as above, it is stated that the operators $+$, $-$, $*$, div and mod have the same semantics as their counterparts within $\mathbb{Z} \times \mathbb{Z}$, when restricted to the set:

$$dom_{Integer} \setminus \{\perp_{Integer}\}$$

For undefined argument values the semantics become:

$$\forall\ x \in dom_{Integer}:$$
$$op\ (x, \perp_{Integer}) = op\ (\perp_{Integer}, x) = \perp_{Integer}$$

where op is a variable copula whose values are respectively the symbols $+$, $-$, $*$, **div** and **mod**.

* The truth table for the null-ary operator $\perp$ is

$$\begin{array}{c|} \perp \\ \hline \perp_{Integer} \end{array}$$

the bottom operation results in an undefined integer value.

$\diamond$

Since each type is defined by a type specification and one or more type implementations, the formal definition of a type can be given as follows:

Definition 1 (Type) *If T represents the set valid of type specifications, P is the infinite set of all type implementations (independent of the language in which they are written) and f_{impl} denotes the mapping function*

$$f_{impl} : T \to \wp(P)$$
$$ts \mapsto \{p_1, \ldots, p_n\}$$

which maps each type specification $ts \in T$ onto the subset $\{p_1, \ldots, p_n\}$ of its co-domain $\wp(P)$ (the powerset of P) that contains all the implementations of ts, then each type is formally defined as a couple:

$$t = [ts, f_{impl}(ts)], \; with \; ts \in T$$

□

Example 2. With the type specification *Integer* being given as in example 1 and the type specification *Boolean* being specified in a similar way:

- $[Integer, f_{impl}(Integer)]$ is the type, which describes the common characteristics of the integer values, whereas
- $[Boolean, f_{impl}(Boolean)]$ is the type, which describes the common characteristics of the Boolean values.

◇

4.2.2 Type system

In order to define the types supported by the presented data model, a type system [16] has been built. The presented type system is consistent with the interface definition language (IDL) specifications of the OMG object model [12,20]. To obtain this consistency, a distinction has been made between so-called basic types and constructed types. (See Fig. 4.2.)

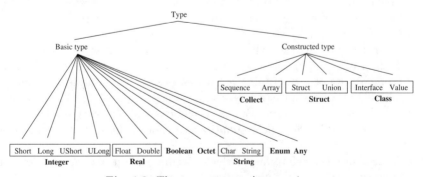

Fig. 4.2. The type system: An overview.

According to the OMG-specifications a basic type is either [12,20]:

- An "Integer"-type; hereby, a distinction is made between 16-bit (Short), 32-bit (Long), and 64-bit signed and unsigned 2's complement integers (UShort and Ulong). Since this distinction is only a matter of physical representation, the integer types will be formally described in the type system by a single type (Integer).
- A "Real"-type; hereby a distinction is made between single-precision 32-bit and fixed-point decimal numbers of up to 31 significant digits (Float), and double precision 64-bit and double extended —a mantissa of at least 64 bits, a sign bit and an exponent of at least 15 bits— IEEE floating point numbers (Double). Since this distinction is also based on different physical representations, a single real type (Real) will be used in the type system to describe the real types.
- A Boolean type (Boolean) taking the values "True" and "False".
- An 8-bit opaque detectable (Octet), guaranteed to not undergo any conversion during transfer between systems.
- A "Text"-type; hereby a distinction is made between (wide) characters, as defined in ISO Latin-1 (8859.1) and other single- or multi-byte character sets (Char), and a string type, which describes variable-length arrays of (wide) characters (String). These types will be formally described in the type system by means of the "String"-type (String).
- An enumerated type (Enum) describing ordered sequences of identifiers.
- A container type "any" (Any), which can represent any possible basic or constructed type.

The basic types are the building blocks of the so-called constructed types. According to the OMG-specifications a constructed type is either [12,20]:

- A generic "Sequence"-type (Sequence), which describe variable-length arrays of a single type.
- A generic "Array"-type (Array), which describe fixed-shape multidimensional arrays of a single type. Both sequence types and array types will be formalized in the type system by means of a general, generic "Collection"-type (Collect).
- A record type (Struct), which describes sets of (name,value) pairs.
- A discriminated union type (Union). Each domain value of an union type consists of a discriminator (whose exact value is always available) followed by an instance of a type appropriate to the discriminator value. Since union types can be seen as special cases of record types, both will be formalized in the type system by means of so-called tuple-types (also denoted as Struct).
- An interface type (Interface), which specifies the set of explicitly defined operations that an instance of that type must support.
- A value type (Value), which specifies structure as well as a set of explicitly defined operations that an instance of that type must support. Since

interface types can be seen as special cases of a value types, both will be formalized in the type system by means of a so-called "object"-type (Class). In accordance with the OMG-specifications, the instances of an object type will be defined as object references. In other words, references to structured values which support a set of explicitly defined operations.

Each type supported by the type system is formally defined as specified in subsection 4.2.1. The syntax rules for the types of the presented type system are defined as follows:

Definition 2 (Syntax rules) *Let ID denote the set of valid identifiers. The set T of expressions that satisfy the syntax of a type and denote the valid type specifications is inductively defined as:*

- *Basic types.*

  - $\{Integer, Real, Boolean, Octet, String, Any\} \subset T.$

  - *If* $id \in ID$ *and* $\{id_1, id_2, \ldots, id_n\} \subseteq ID$, *then*
    $$Enum\ id(id_1, \ldots, id_n) \in T \text{ (enumeration type)}.$$

    The identifier $id \in ID$ *identifies the enumeration type, whereas* $(id_1, id_2, \ldots, id_n)$ *represents the ordered sequence of identifiers, which is described by the type.*

- *Constructed types.*

  - *If* $t \in T \setminus \{Any\}$, *then* $Collect(ts) \in T$ *(collection type).*

  - *If* $id \in ID$, $\{id_1, id_2, \ldots, id_n\} \subseteq ID$ *and* $t_i \in T \setminus \{Any\}$, $1 \leq i \leq n$, *then* $Struct\ id(id_1 : t_1;\ id_2 : t_2;\ \ldots;\ id_n : t_n) \in T$ *(tuple-type).*

    The identifier id identifies the tuple-type, whereas $(id_1 : t_1, id_2 : t_2, \ldots, id_n : t_n)$ *represents the components of the tuple-type. Each component* $id_i : t_i$, $1 \leq i \leq n$ *is characterized by an identifier* id_i *and a type* t_i.

  - *Let* V_{signat} *denote the set of all valid operation signatures, which is defined as follows:*
    * *If* $t' \in T$ *then* $Signat(( ) \rightarrow t') \in V_{signat}.$
    * *If* $t' \in T$, $\{id'_1, id'_2, \ldots, id'_p\} \subseteq ID$ *and* $t'_i \in T$, $1 \leq i \leq p$, *then*
      $$Signat((id'_1 : t'_1;\ id'_2 : t'_2;\ \ldots;\ id'_p : t'_p) \rightarrow t') \in$$
      $V_{signat}.$

Hereby, id'_i and t'_i, $1 \le i \le p$ are respectively the names and the expected types of the input parameters of the operation and t' is the type of the returned value(s) of the operation.

If $id \in ID$, $\{\widehat{id}_1, \widehat{id}_2, \ldots, \widehat{id}_m\} \subset ID \setminus \{id\}$, $\{id_1, id_2, \ldots, id_n\} \subset ID$ and $s_i \in T \cup V_{signat} \setminus \{Any\}$, $1 \le i \le n$, then

* *Class $id(id_1 : s_1; \ id_2 : s_2; \ \ldots; \ id_n : s_n) \in T$*

* *Class $id : \widehat{id}_1, \widehat{id}_2, \ldots, \widehat{id}_m(\ ) \in T$*

* *Class $id : \widehat{id}_1, \widehat{id}_2, \ldots, \widehat{id}_m(id_1 : s_1; \ id_2 : s_2; \ \ldots; \ id_n : s_n) \in T$*
(object type)

The identifier id identifies the object type, the identifiers $\widehat{id}_i$, $1 \le i \le m$ are the identifiers of the supertypes of the object type, whereas $(id_1 : s_1, id_2 : s_2, \ldots, id_n : s_n)$ represents the characteristics[1] of the object type. (Beside these characteristics, the object type can inherit other characteristics from its parent types.)

□

Furthermore, the full semantics of the types with specification $ts \in T$ (cf. definition 2) are defined by providing an appropriate definition for the set of domains D_{ts}, the domain dom_{ts} of the type, the set of operators O_{ts} and the set of axioms A_{ts}. Full definitions are beyond the scope of this chapter. Instead an informal description is given:

- *Basic types.* The definition of the basic types is straightforward. Each basic type has a domain, which consists of simple —non-composed— values. Its corresponding set of operators consists of the usual operators defined over its domain.
  - The domain of the *Integer*-type consists of integer numbers and the "undefined" value $\perp_{Integer}$. The set of operators $O_{Integer}$ consists of the operators $\perp, =, \ne, <, >, \le, \ge, +, -, *, div$ and mod (cf. example 1).

  - The domain of the *Real*-type consists of real numbers and the "undefined" value $\perp_{Real}$, whereas O_{Real} consists of the operators $\perp, =, \ne, <, >, \le, \ge, +, -, *$ and $/$.

  - The domain of the *Boolean*-type is the set with the "undefined" value $\perp_{Boolean}$ and the truth values $TRUE$ and $FALSE$. The set of operators $O_{Boolean}$ consists of the operators $\perp, =, \ne, \wedge, \vee$ and $\neg$.

---

[1] The term characteristic is used to denote both the attributes (structure) and the operations (behavior) of the object type.

– The domain of the *Octet*-type consists of 8-bit binary values (bytes) and the "undefined" value $\perp_{Octet}$. The set of operators O_{Octet} consists of the operators $\perp$, $=$, $\neq$, $<$, $>$, $\leq$, $\geq$, $\wedge$, $\vee$, $\neg$, *Shift_left* and *Shift_right*.

– The domain of the *String*-type consists of character strings and the "undefined" value $\perp_{String}$, whereas O_{String} consists of the operators $\perp$, $=$, $\neq$, $<$, $>$, $\leq$, $\geq$, *concat*, *sizeof* and *substr*.

– The domain of an enumeration type with specification *ts* consists of the undefined domain value $\perp_{ts}$ and the identifiers specified in the type specification. Its corresponding set of operators consists of the operators $\perp$, $=$, $\neq$, $<$, $>$, $\leq$ and $\geq$.

– Finally, the domain of the *Any*-type is by definition the singleton $\{\perp_{Any}\}$. Its corresponding set of operators is the singleton

$$\{\perp : \rightarrow dom_{Any}\}$$

• *Constructed types.* Constructed types are characterized by composed domain values.
  – *Collection type.* The collection type is a generic type, indicated by the type generator *Collect* and a type parameter t. The domain of a collection type *Collect*(t) consists of the "undefined" domain value $\perp_{Collect(t)}$ and of sets of elements of the domain of type t. The associated set of operators consists of the operators $\perp$, $=$, $\neq$, $\subset$, $\supset$, $\subseteq$, $\supseteq$, *cardinality*, *is_empty*, $\cup$, $\cap$, $\setminus$ and *contains_element*.

  – *Tuple-type.* A tuple-type is structured and consists of a fixed number of components $id_i : ts_i$, $i = 1, 2, \ldots, n$. Each component $id_i : ts_i$ consists of a unique identifier id_i and a type specification ts_i. The domain of a tuple-type with specification *ts* and n components $id_i : ts_i$, $i = 1, 2, \ldots, n$ contains the "undefined" domain value $\perp_{ts}$. All other domain values are composed and consist of n values $id_i : v_i$, with $v_i \in dom_{ts_i}$, $i = 1, 2, \ldots, n$. The associated set of operators consists of the operators $\perp$, $=$, $\neq$, *set_component* and *get_component*.

  – *Object type.* The object types are the most elaborated types of the type system. Each object type is characterized by a number of attributes (which describe its structure) and a number of explicitly defined operators —also called methods— (which describe its behavior). An object type can inherit attributes and methods from its parent types. In order to define the inheritance-based type-subtype relationships between object types, a partial ordering function $\prec$ is defined over the set T of type specifications. ($\widehat{id} \prec id$ denotes that "object type id inherits all characteristics of object type $\widehat{id}$".)

The domain of an object type with specification *ts* contains the "undefined" domain value $\perp_{ts}$. Every other instance is a reference to an object of that type or a reference to an object of one of its subtypes. Each object is composed and contains a value $id : v$, for each of the (inherited) attributes $id : s$, $s \in T$ of its type. Hereby, $v \in dom_s$. The set of operators associated with a given object type is the union of a set of implicitly defined operators and a set of explicitly defined operators. The implicitly defined operators are: $\perp$, $=$, $\neq$, *set_attribute* and *get_attribute*. The explicitly defined operators are the (inherited) methods $id : s$, $s \in V_{signat}$ of the object type.

The type system *TS*, which defines all the valid types supported by the presented data model, is defined as:

Definition 3 (Type system) *The type system TS is formally defined by the quintuple*

$$TS = [ID, T, P, f_{impl}, \prec]$$

where

- *ID is the set of the valid identifiers,*
- *T is the set of valid type specifications (cf. definition 2),*
- *P is the infinite set of all type implementations (cf. definition 1),*
- *f_{impl} is the mapping function, which maps each type specification onto its corresponding set of type implementations (cf. definition 1) and*
- *$\prec : T^2 \rightarrow \{True, False\}$ is the partial ordering which is used to define the inheritance-based type-subtype relationships between object types.*

□

Example 3. The type system allows for definitions like the following, which are intended to describe a (simplified) type for employees. With the tuple-types with specification

Struct TAddress(Street : String; City : String; Zip : Integer),

Struct TCompany(Name : String; Location : String)

and

Struct TWorks(Company : TCompany; Percentage : Real)

and the enumeration type with specification

Enum TLang(Dutch, French, English, German, Spanish, Italian) the specifications of the object types

$$[TPerson, f_{impl}(TPerson)] \text{ and } [TEmployee, f_{impl}(TEmployee)]$$

can be defined as:

> *Class TPerson(*
> > *Name : String;*
> > *Age : Integer;*
> > *Address : TAddress;*
> > *Languages : Collect(TLang);*
> > *Children : Collect(TPerson);*
> > *Add_child : Signat ((New_child : TPerson) → Any))*

and

> > *Class TEmployee : TPerson(*
> > > *EmployeeID : String;*
> > > *Works_for : Collect(TWorks))*

◇

Example 4. The instances of the object type *TPerson* of example 3 are either references to "TPerson"-objects or references to "TEmployee"-objects (since *TEmployee* is a subtype of *TPerson*). The instances of the object type *TEmployee* must be references to "TEmployee"-objects.

Examples of "TPerson"-objects are:

> *(Name : "Ann";*
> *Age : 14;*
> *Address : (Street : "ChurchStreet, 12"; City : "Brussels"; Zip : 1000);*
> *Languages : (Dutch, French);*
> *Children : ())*

and

> *(Name : "Joe";*
> *Age : 18;*
> *Address : (Street : "ChurchStreet, 12"; City : "Brussels"; Zip : 1000);*
> *Languages : (Dutch, French, English);*
> *Children : ())*

An example of a "TEmployee"-object is:

> (*Name* : *"John"*;
>   *Age* : 42;
>   *Address* : (*Street* : *"ChurchStreet*, 12"; *City* : *"Brussels"*; *Zip* : 1000);
>   *Languages* : (*French*, *English*, *Dutch*, *Italian*);
>   *Children* : (*ref_to_Ann*, *ref_to_Joe*);
>   *EmployeeID* : *"ID254"*;
>   *Works_for* : ((*Company* : (*Name* : *"IBM"*; *Location* : *"Brussels"*);
>   *Percentage* : 100)))

◇

4.3   Generalized Types and Generalized Type System

The concept "generalized type" will be defined as a generalization of the concept "type". It can informally be seen as a description of the common characteristics of a collection of data in its most general form (i.e. crisp, uncertain, imprecise, vague, incomplete, and any valid combination of these). Generalized types provide a more adequate means to model "real-world" entities, since they allow to model the characteristics of an entity in a more natural way. The definition of a generalized type is given in subsection 4.3.1.

In order to have a formalism for the uniform handling of both crisply and non-crisply described information, the data model described by the type system of subsection 4.2.2 is generalized. Hereby, a so-called generalized type system is defined (cf. subsection 4.3.2). In general, a generalized type system can be seen as a collection of rules, which together define the generalized types supported by a given model [6]. The presented generalized type system describes in a formal way how fuzziness and/or uncertainty about the data could be handled in any object-oriented software engineering or data engineering application.

4.3.1   Generalized types

Each generalized type is a type and thus defined by a type specification and one or more type implementations. The type specification determines the syntax and the semantics of the generalized type. The semantic definition of a generalized type $\tilde{t}$ is obtained as a generalization of the semantic definition of the type t, for which $\tilde{t}$ is the generalization. Central to this generalization is the introduction of the concept of a "generalized domain" and the impact of this introduction on the definition of the operations of a type.

Generalized domains. The definition of a generalized domain is based on the concept of a level-2 fuzzy set, i.e. a fuzzy set which is defined over a universe of (ordinary) fuzzy sets [11,13]. If U represents the universe over which these ordinary fuzzy sets are defined and $\tilde{\wp}(U)$ denotes the set of all the fuzzy sets, which can be defined on the universe U, then a level-2 fuzzy set $\tilde{\tilde{V}}$ is defined by:

$$\tilde{\tilde{V}} = \{(\tilde{V}, \mu_{\tilde{\tilde{V}}}(\tilde{V})) \mid \forall \ \tilde{V} \in \tilde{\wp}(U) : \mu_{\tilde{\tilde{V}}}(\tilde{V}) > 0\}$$

where each ordinary fuzzy set $\tilde{V}$ is defined by:

$$\tilde{V} = \{(x, \mu_{\tilde{V}}(x)) \mid \forall \ x \in U : \mu_{\tilde{V}}(x) > 0\}$$

The level-2 fuzzy set $\tilde{\tilde{V}}$ is shortly called a level-2 fuzzy set, which is defined over the universe U.

Example 5. The expression

$$\{(\{(a, 1), (b, 1)\}, 0.3), (\{(c, 0.8), (d, 1), (e, 0.8)\}, 0.9)\}$$

is a level-2 fuzzy set, which consists of two ordinary fuzzy sets $\{(a, 1), (b, 1)\}$ (with membership degree 0.3) and $\{(c, 0.8), (d, 1), (e, 0.8)\}$ (with membership degree 0.9).
◇

With respect to the generalization of a domain, the membership degrees of a level-2 fuzzy set are interpreted as follows [8]: (See Fig. 4.3.)

- The "outer-layer" membership degrees $\mu_{\tilde{\tilde{V}}}(\tilde{V})$ of the fuzzy sets $\tilde{V} \in \tilde{\wp}(U)$ of a given level-2 fuzzy set $\tilde{\tilde{V}}$ are always interpreted as degrees of uncertainty and are used for the modeling of uncertain information.

- The "inner-layer" membership degrees $\mu_{\tilde{V}}(x)$ of the elements $x \in U$ in a given level-2 fuzzy set $\tilde{\tilde{V}}$ are either all interpreted as degrees of similarity or all interpreted as degrees of preference —depending on the context in which the level-2 fuzzy set is applied— and are used for the modeling of fuzzy information (i.e. imprecise information or vague information).

Modeling information by means of level-2 fuzzy sets, as introduced above, allows the user to perceive clearly the difference between the fuzziness and the uncertainty, as if they pertain to two nested layers, the "inner layer" corresponding to the fuzzy description of the data themselves, the "outer layer" to the (un)certainty about the existence of the data. Moreover uncertainty, imprecision, vagueness and any valid combination of these can be modeled in a uniform way.

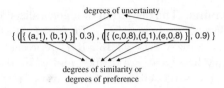

degrees of uncertainty

{ ({ (a,1), (b,1) }, 0.3) , ({ (c,0.8),(d,1),(e,0.8) }, 0.9) }

degrees of similarity or
degrees of preference

Fig. 4.3. The interpretation of the membership degrees of a level-2 fuzzy set.

Example 6. The interpretation of the level-2 fuzzy set

$$\{(\{(a,1),(b,1)\},0.3),(\{(c,0.8),(d,1),(e,0.8)\},0.9)\}$$

is as follows:

- If the "inner-layer" membership degrees are all interpreted as degrees of similarity, then the level-2 fuzzy set represents a domain value, which is less possibly —with uncertain degree 0.3— either a or b or more possibly —with uncertain degree 0.9— either c, d or e. (*Disjunctive interpretation* of the "inner-layer" fuzzy sets.)

  As an example consider the level-2 fuzzy sets defined over the set of integer values. With a disjunctive interpretation of the "inner-layer" membership degrees, these level-2 fuzzy sets are suitable to represent the age of a person. For example the level-2 fuzzy set

  $$\{(\{(25,1),(26,1)\},0.3),(\{(44,0.8),(45,1),(46,0.8)\},0.9)\}$$

  represents the age of a person who is either 25 or 26 years old or is either around 45 years old.

- If the "inner-layer" membership degrees are all interpreted as degrees of preference, then the level-2 fuzzy set represents a domain value, which is less possibly —with uncertain degree 0.3— a and b or more possibly —with uncertain degree 0.9— c, d and e. (*Conjunctive interpretation* of the "inner-layer" fuzzy sets.)

  For example, consider the level-2 fuzzy sets defined over the set of natural languages. With a conjunctive interpretation of the "inner-layer" membership degrees, these level-2 fuzzy sets are suitable to represent the languages spoken by a person. For example the level-2 fuzzy set

  $$\{(\{(English,1)\},0.3),(\{(English,1),(Dutch,0.7)\},0.9)\}$$

  represents that the person either has only a fluent knowledge of English, or either has a fluent knowledge of English and and a less fluent knowledge of Dutch.

◇

In this setting "crisp" values $x \in U$ can still be modeled by means of a normalized level-2 fuzzy set, which consists of a single normalized fuzzy set, which on its turn consists of the single "crisp" value, i.e. as a level-2 fuzzy set

$$\{(\{(x,1)\},1)\}, \text{ where } x \in U$$

The domain $dom_{\tilde{ts}}$ of a generalized type with specification $\tilde{ts}$ has been defined as the set of all the level-2 fuzzy sets, which can "meaningfully" be defined over the domain dom_{ts} of the corresponding type with specification ts. Therefore, two possible cases have to be considered:

- *Case 1: It is meaningful to consider both uncertainty and fuzziness with respect to the values of the domain dom_{ts}.*
  In this case the generalized domain $dom_{\tilde{ts}}$ is defined by:

  $$dom_{\tilde{ts}} = \tilde{\wp}(\tilde{\wp}(dom_{ts}))$$

  where $\tilde{\wp}(U)$ denotes the set of all the fuzzy sets, which can be defined on the universe U. This is the most general way to generalize a domain dom_{ts}. The domain $dom_{Integer}$ of the integer type (cf. example 1) is an example of a domain, which can be generalized in this way.

- *Case 2: It is only meaningful to consider uncertainty with respect to the values of the domain dom_{ts}.*
  In this case the generalized domain $dom_{\tilde{t}}$ is defined by:

  $$dom_{\tilde{ts}} = \tilde{\wp}(\{\{(x,1)\} \mid x \in dom_{ts}\})$$

  where $\tilde{\wp}(U)$ denotes the set of all the fuzzy sets, which can be defined on the universe U. Logically, with this definition, no imprecise or vague values are allowed. The domain $dom_{Boolean}$ of the Boolean type is an example of a domain which can be generalized in this way (because vague or imprecise Boolean values can be considered as not being meaningful —but uncertainty about a Boolean value is still allowed—).

With respect to a given type with specification ts, all the "inner-layer" fuzzy sets $\tilde{V} \in \tilde{\wp}(dom_{ts})$, which contain the "undefined" domain value $\perp_{ts}$ (i.e. for which, $\mu_{\tilde{V}}(\perp_{ts}) > 0$) have been defined to be equal. This is inspired by the fact that it does not make sense to consider imprecision with respect to an "undefined" domain value, as should for example be the case in "*approximately undefined*". Therefore, Zadeh's standard definition for equality [21] is extended as follows:

Definition 4 (Equality of fuzzy sets – extended)

$$\forall \tilde{U}, \tilde{V} \in \tilde{\wp}(dom_{ts}) : \tilde{U} = \tilde{V} \Leftrightarrow$$
$$(\forall x \in dom_{ts} : \mu_{\tilde{U}}(x) = \mu_{\tilde{V}}(x)) \vee (\mu_{\tilde{U}}(\perp_{ts}) > 0 \wedge \mu_{\tilde{V}}(\perp_{ts}) > 0)$$

□

Uncertainty can still occur in combination with an "undefined" domain value as it is the case in e.g. "John's partner is *possibly Mary, possibly Jane or —less possibly— John has no partner at all*", which can be modeled as

$$\{(\{(Mary, 1)\}, 1), (\{(Jane, 1)\}, 1), (\{(\perp_{Name}, 1)\}, .3)\}$$

With the previous generalization of the definition of the domain of a type all domain values are level-2 fuzzy sets, which allow for the modeling of crisp information, as well as uncertain information, vague information, imprecise information and uncertainty about vague or imprecise information.

Example 7. Reconsider the modeling of the age of a person. The domain $dom_{Integer}$ of the integer type (cf. example 1) can be generalized as follows:

$$dom_{Integer}^{\tilde{z}} = \tilde{\wp}(\tilde{\wp}(dom_{Integer}))$$

The values of the resulting generalized domain $dom_{Integer}^{\tilde{z}}$ can be used to model information about the age of a person, which is either given

- by means of a crisp number; e.g. "20 years old" is modeled by the generalized domain value

$$\{(\{(20, 1)\}, 1)\}$$

- in an imprecise way; e.g. "between 15 and 20 years old", can be represented by the generalized domain value

$$\{(\tilde{V}_{between\ 15\ and\ 20}, 1)\}$$

with $\tilde{V}_{between\ 15\ and\ 20}$ a disjunctive fuzzy set, which represents an age "between 15 and 20 years old";
- as vague information, described by means of a linguistic term; e.g. "young", can be represented by the generalized domain value

$$\{(\tilde{V}_{young}, 1)\}$$

with $\tilde{V}_{young}$ a disjunctive fuzzy set, which represents a young age;
- in an uncertain way; e.g. "possibly 20 years old", can be represented by the generalized domain value

$$\{(\{(19, 1)\}, .8), (\{(20, 1)\}, 1), (\{(21, 1)\}, .8)\}$$

- in an imprecise or vague way, combined with uncertainty, e.g. "possibly between 25 and 30 years old, less possibly middle aged", can be represented by the generalized domain value

$$\{(\tilde{V}_{between\ 25\ and\ 30}, .8), (\tilde{V}_{middle\_aged}, .4)\}$$

where the disjunctive fuzzy sets $\tilde{V}_{between\ 25\ and\ 30}$ and $\tilde{V}_{middle\_aged}$ respectively represent an age "between 25 and 30 years old" and a "middle_aged" age.

◇

Generalized operations. The generalization of the domain of a type —as proposed in the previous subsection— has a direct impact on the definition of the operators of the generalized counterpart of this type. With respect to (the generalized domain of) a generalized type with specification $\tilde{\tilde{ts}}$, three kinds of operators are distinguished: the so-called extended operators, generalized operators and specific operators. This is illustrated in Fig. 4.4.

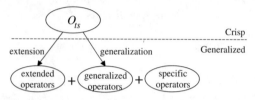

Fig. 4.4. The operators of a generalized type.

1. *Extended operators.* Each extended operator $\tilde{\tilde{\Phi}}$ is by definition an extension of an operator Φ of the corresponding type with specification ts (see Fig. 4.4). If $\Phi \in O_{ts}$ is specified as

$$\Phi : dom_{ts_1} \times dom_{ts_2} \times \cdots \times dom_{ts_n} \to dom_{ts_{n+1}}$$

then $\tilde{\tilde{\Phi}}$ is specified as

$$\tilde{\tilde{\Phi}} : dom_{\tilde{\tilde{ts}}_1} \times dom_{\tilde{\tilde{ts}}_2} \times \cdots \times dom_{\tilde{\tilde{ts}}_n} \to dom_{\tilde{\tilde{ts}}_{n+1}}$$

If all the operands of $\tilde{\tilde{\Phi}}$ represent crisp domain values, then $\tilde{\tilde{\Phi}}$ has (by definition) the same behavior as Φ. If at least one of these operands represents a non-crisp value, then $\tilde{\tilde{\Phi}}$ results in the "undefined" generalized domain value $\{(\{(\perp_{ts_{n+1}}, 1)\}, 1)\}$, i.e.

$$\forall \tilde{\tilde{x}}_1, \ldots, \tilde{\tilde{x}}_n \in dom_{\tilde{\tilde{ts}}_1} \times \cdots \times dom_{\tilde{\tilde{ts}}_n} : \tilde{\tilde{\Phi}}(\tilde{\tilde{x}}_1, \ldots, \tilde{\tilde{x}}_n) =$$

$$\begin{cases} f_t^{-1}(\Phi(f_t(\tilde{\tilde{x}}_1, \ldots, \tilde{\tilde{x}}_n))) & \text{if } \forall\ i = 1, 2, \ldots, n, \exists\ x_i \in dom_{ts_i} : \\ & \qquad\qquad \tilde{\tilde{x}}_i = \{(\{(x_i, 1)\}, 1)\} \\ \{(\{(\perp_{ts_{n+1}}, 1)\}, 1)\} & \text{in all other cases} \end{cases}$$

where the transformation function f_t is defined by

$$\forall \tilde{\tilde{x}}_1, \ldots, \tilde{\tilde{x}}_n \in dom_{\tilde{\tilde{ts}}_1} \times \cdots \times dom_{\tilde{\tilde{ts}}_n} : f_t((\tilde{\tilde{x}}_1, \ldots, \tilde{\tilde{x}}_n)) = (x_1', \ldots, x_n')$$

with

$$\forall\ i = 1, 2, \ldots, n : x_i' = \begin{cases} x_i & \text{if } \tilde{\tilde{x}}_i = \{(\{(x_i, 1)\}, 1)\} \\ \perp_{ts_i} & \text{in all other cases} \end{cases}$$

and the "inverse" transformation function f_t^{-1} is defined by

$$f_t^{-1} : dom_{ts_{n+1}} \to dom_{\tilde{\tilde{ts}}_{n+1}}$$

$$x_{n+1} \mapsto \{(\{((x_{n+1}, 1)\}, 1)\}$$

2. *Generalized operators.* Each generalized operator $\tilde{\tilde{\Phi}}$ is by definition a generalization of an operator Φ of the corresponding type with specification ts (see Fig. 4.4). If $\Phi \in O_{ts}$ is specified as

$$\Phi : dom_{ts_1} \times dom_{ts_2} \times \cdots \times dom_{ts_n} \to dom_{ts_{n+1}}$$

then $\tilde{\tilde{\Phi}}$ is specified as

$$\tilde{\tilde{\Phi}} : dom_{\tilde{\tilde{ts}}_1} \times dom_{\tilde{\tilde{ts}}_2} \times \cdots \times dom_{\tilde{\tilde{ts}}_n} \to dom_{\tilde{\tilde{ts}}_{n+1}}$$

and obtained by applying Zadeh's extension principle two successive times to Φ. Two cases have been distinguished, depending on the way $dom_{\tilde{\tilde{ts}}_{n+1}}$ is defined. Let f_{ext} be the function, which maps each crisp relation

$$R : U_1 \times U_2 \times \cdots \times U_n \to Y$$

onto the fuzzy relation

$$\tilde{R} : \tilde{\wp}(U_1) \times \tilde{\wp}(U_2) \times \cdots \times \tilde{\wp}(U_n) \to \tilde{Y}$$

which is obtained by applying Zadeh's extension principle [22].

(a) If the domain $dom_{\tilde{\tilde{ts}}_{n+1}}$ is defined by

$$dom_{\tilde{\tilde{ts}}_{n+1}} = \tilde{\wp}(\tilde{\wp}(dom_{ts_{n+1}}))$$

then the generalized operator $\tilde{\tilde{\Phi}}$ is by definition

$$\tilde{\tilde{\Phi}} = f_{ext}(f_{ext}(\Phi))$$

(b) If the domain $dom_{\tilde{\tilde{ts}}_{n+1}}$ is defined by

$$dom_{\tilde{\tilde{ts}}_{n+1}} = \tilde{\wp}(\{\{(x, 1)\} \mid x \in dom_{ts_{n+1}}\})$$

then the generalized operator $\tilde{\tilde{\Phi}}$ is by definition

$$\tilde{\tilde{\Phi}} = f_t(f_{ext}(f_{ext}(\Phi)))$$

where f_t is the transformation function, which transforms each level-2 fuzzy set

$$\tilde{\tilde{X}} \in \tilde{\wp}(\tilde{\wp}(dom_{ts_{n+1}}))$$

into its corresponding level-2 fuzzy set

$$f_t(\tilde{\tilde{X}}) \in \tilde{\wp}(\{\{(x,1)\} \mid x \in dom_{ts_{n+1}}\})$$

which is defined by the membership function

$$\mu_{f_t} : \{\{(x,1)\} \mid x \in dom_{ts_{n+1}}\} \to [0,1]$$

$$\{(x,1)\} \mapsto \sup_{\tilde{X} \in \{\tilde{X} \mid \tilde{X} \in \tilde{\wp}(dom_{ts_{n+1}}) \, \wedge \, \mu_{\tilde{\tilde{X}}}(\tilde{X}) \neq 0\}} min(\mu_{\tilde{\tilde{X}}}(\tilde{X}), \mu_{\tilde{X}}(x))$$

The application of the transformation function f_t guarantees that the generalized operator $\tilde{\tilde{\Phi}}$ always results in an element of its co-domain $dom_{\tilde{\tilde{ts}}_{n+1}}$.

3. *Specific operators.* Because by definition all the elements of a generalized domain are level-2 fuzzy sets —and are a fortiori fuzzy sets—, it is useful to provide extra operators for the handling of level-2 fuzzy sets and their "inner-level" fuzzy sets. These operators are called "specific" operators (see Fig. 4.4). Examples of specific operators are the operators $\cup$, $\cap$, *co*, *normalize*, *support*, *core*, $\alpha - cut$ and $\bar{\alpha} - cut$. Each specific operator preserves its usual semantics.

Definition of generalized types. With the generalizations of domains and operators as described in the previous subsections and analogous to the specification ts of a type, the specification $\tilde{\tilde{ts}}$ of a generalized type can be determined by:

- *The syntax of the generalized type (specification):* The syntax rules for a generalized type specification can be formally described by means of some mathematical expressions.
- *The semantics of the generalized type (specification):* The semantic definition of a generalized type with specification $\tilde{\tilde{ts}}$ can be fully determined by:
  - a set of generalized domains $D_{\tilde{\tilde{ts}}}$,
  - a designated generalized domain $dom_{\tilde{\tilde{ts}}} \in D_{\tilde{\tilde{ts}}}$,
  - a set of operators $O_{\tilde{\tilde{ts}}}$ and
  - a set of axioms $A_{\tilde{\tilde{ts}}}$

Analogous to the semantic definition of a type, the designated domain $dom_{\tilde{\tilde{ts}}}$ defines the set of all the possible values for $\tilde{\tilde{ts}}$; the set of operators $O_{\tilde{\tilde{ts}}}$ contains the operators that are defined on the generalized domain

$dom_{\tilde{ts}}$; the set of generalized domains $D_{\tilde{ts}}$ consists of the generalized domains, which are involved in the definition of the operators of $O_{\tilde{ts}}$; whereas the set of axioms $A_{\tilde{ts}}$ contains the axioms, which are involved in the definition of the semantics of $O_{\tilde{ts}}$.

Example 8. The generalized type specification, which describes the common characteristics of the (generalized) integer values, i.e. the generalization of the "Integer"-type (cf. example 1), is fully determined as follows:

- *Syntax.* The generalized type for the modeling of integer values is represented by the symbol $\widetilde{Integer}$.
- *Semantics.* The semantics of the generalized type $\widetilde{Integer}$ have been defined by:
  - The designated generalized domain $dom_{\widetilde{Integer}}$:

  $$dom_{\widetilde{Integer}} = \tilde{\wp}(\tilde{\wp}(dom_{Integer}))$$

  - The set of generalized domains $D_{\widetilde{Integer}}$:

  $$D_{\widetilde{Integer}} = \{dom_{\widetilde{Boolean}}, dom_{\widetilde{Integer}}\}$$

  - The set of operators $O_{\widetilde{Integer}}$:

  $$O_{\widetilde{Integer}} = \{ \widetilde{=} : (dom_{\widetilde{Integer}})^2 \to dom_{\widetilde{Boolean}},$$
  $$\widetilde{\neq} : (dom_{\widetilde{Integer}})^2 \to dom_{\widetilde{Boolean}},$$
  $$\widetilde{<} : (dom_{\widetilde{Integer}})^2 \to dom_{\widetilde{Boolean}},$$
  $$\widetilde{>} : (dom_{\widetilde{Integer}})^2 \to dom_{\widetilde{Boolean}},$$
  $$\widetilde{\leq} : (dom_{\widetilde{Integer}})^2 \to dom_{\widetilde{Boolean}},$$
  $$\widetilde{\geq} : (dom_{\widetilde{Integer}})^2 \to dom_{\widetilde{Boolean}},$$
  $$\widetilde{+} : (dom_{\widetilde{Integer}})^2 \to dom_{\widetilde{Integer}},$$
  $$\widetilde{-} : (dom_{\widetilde{Integer}})^2 \to dom_{\widetilde{Integer}},$$

  $$\widetilde{*} : (dom_{\widetilde{Integer}})^2 \to dom_{\widetilde{Integer}},$$
  $$\widetilde{div} : (dom_{\widetilde{Integer}})^2 \to dom_{\widetilde{Integer}},$$
  $$\widetilde{mod} : (dom_{\widetilde{Integer}})^2 \to dom_{\widetilde{Integer}},$$
  $$\perp : \to dom_{\widetilde{Integer}}\} \cup O^{specific}$$

where $O^{specific}$ is the set of the specific operators, which are provided for the handling of (level-2) fuzzy sets.

- The set of axioms $A_{\underset{\sim}{Integer}}$:
  * The semantics of the binary operators $\overset{\approx}{=}$, $\overset{\approx}{\neq}$, $\overset{\approx}{<}$, $\overset{\approx}{>}$, $\overset{\approx}{\leq}$, $\overset{\approx}{\geq}$, $\overset{\approx}{+}$, $\overset{\approx}{-}$, $\overset{\approx}{*}$, $\overset{\approx}{\text{div}}$ and $\overset{\approx}{\text{mod}}$:

    the operators $\overset{\approx}{=}$, $\overset{\approx}{\neq}$, $\overset{\approx}{<}$, $\overset{\approx}{>}$, $\overset{\approx}{\leq}$, $\overset{\approx}{\geq}$, $\overset{\approx}{+}$, $\overset{\approx}{-}$, $\overset{\approx}{*}$, $\overset{\approx}{\text{div}}$ and $\overset{\approx}{\text{mod}}$ are defined as generalized operators. Therefore, their full semantics are completely defined by the semantics of their "crisp" counterparts in the set $O_{Integer}$ and the definition of the generalization (as given in the previous subsection).

  * The semantics of the operators of the set $O^{specific}$:
    each specific operator preserves its usual semantics, as stated in the previous subsection.

  * The truth table for the null-ary operator $\bot$:

$$\begin{array}{c|c} \bot & \\ \hline & \{(\{(\bot_{Integer}, 1)\}, 1)\} \end{array}$$

    the bottom operation always results in an "undefined" generalized integer value.

◇

Since each generalized type is defined by a generalized type specification and one or more type implementations, the formal definition of a generalized type is given as follows:

Definition 5 (Generalized type) *If $\overset{\approx}{T}$ represents the set valid of generalized type specifications, P is the infinite set of all type implementations and $\overset{\approx}{f}_{impl}$ denotes the mapping function*

$$\overset{\approx}{f}_{impl} : \overset{\approx}{T} \to \wp(P)$$
$$\overset{\approx}{ts} \mapsto \{p_1, \ldots, p_n\}$$

which maps each generalized type specification $\overset{\approx}{ts} \in \overset{\approx}{T}$ onto the subset $\{p_1, \ldots, p_n\}$ of its co-domain $\wp(P)$ that contains all the implementations of $\overset{\approx}{ts}$, then each generalized type is formally defined as a couple:

$$\overset{\approx}{t} = [\overset{\approx}{ts}, \overset{\approx}{f}_{impl}(\overset{\approx}{ts})], \text{ with } \overset{\approx}{ts} \in \overset{\approx}{T}$$

□

Example 9. With the generalized type specification $\overset{\approx}{Integer}$ being given as in example 8, the generalized type which describes the common characteristics of "generalized" integer values is

$$[\overset{\approx}{Integer}, \overset{\approx}{f}_{impl}(\overset{\approx}{Integer})]$$

◇

4.3.2   Generalized type system

In order to obtain a generalized data model, which can be used to handle both crisply described information and non-crisply described information, the type system

$$TS = [ID, T, P, f_{impl}, \prec]$$

of subsection 4.2.2 has been generalized. The types supported by the generalized type system are the generalizations of the types specified by the type system TS.

First of all, the set of type specifications T has been generalized to the set $\tilde{T}$ by generalizing the definitions of the domains and the operators of each type specification $\tilde{ts} \in \tilde{T}$. The basic types and collection types are generalized as discussed in subsection 4.3.1. Tuple-types are generalized by generalizing (the definition of) their components, whereas for object types the generalization is obtained by generalizing (the definition of) their characteristics.

The generalization of components and characteristics has been realized by using generalized constraints as defined by L.A. Zadeh [26,5]. Each component and characteristic

$$id : s$$

is generalized as:

$$id \; isr \; \tilde{s}$$

where $\tilde{s}$ is the generalized counterpart of the specification s of the component or characteristic and isr is a variable copula with r a discrete variable, which value defines the way in which the values of $\tilde{s}$ are constrained. Two cases are considered:

1. If $id : s$ is a component of a structured type or an attribute of an object type, then s has to be a type specification. The variable copula isr then denotes the generalized constraint, which constrains the valid domain values of the generalized counterpart of s, i.e. the valid domain values of the generalized type with specification $\tilde{s}$. Furthermore, the generalized constraint defines the interpretation of the "inner-layer" membership degrees, which appear in these values. For now, the considered types of generalized constraints are:
   - *Equality constraint,* $r = e$. In this case, $id \; ise \; \tilde{t}$ means that id must be assigned a precise and accurately described value of $dom_{\tilde{t}}$, i.e. a level-2 fuzzy set $\tilde{V} \in \wp(\{\{(v,1)\} \mid v \in dom_t\})$ which represents either certain or uncertain information that is crisply described. This case is almost semantically equivalent to the definition $id : t$ in the crisp counterpart, except that it allows the data to be uncertain too.

- *Possibilistic constraint, $r = blank$.* In this case, *id is $\tilde{\tilde{t}}$* means that *id* must be assigned a value of $dom_{\tilde{\tilde{t}}}$, i.e. a level-2 fuzzy set, for which the "inner-layer" membership degrees are interpreted as degrees of similarity [8]. The component with identifier *id* is interpreted as a disjunctive (possibilistic) variable.
- *Veristic constraint, $r = v$.* In this case, *id isv $\tilde{\tilde{t}}$* means that *id* must be assigned a value of $dom_{\tilde{\tilde{t}}}$, i.e. a level-2 fuzzy set, for which the "inner-layer" membership degrees are interpreted as degrees of preference [8]. The component with identifier *id* is interpreted as a conjunctive (veristic) variable.

2. If $id : s$ is an operation, then s has to be an operation signature. The signature $\tilde{\tilde{s}}$ of the generalized counterpart of the operation is then obtained as follows:

- If $s = Signat((\ ) \to t')$, then $\tilde{\tilde{s}} = Signat((\ ) \to \tilde{\tilde{t}}')$

- If $s = Signat((id'_1 : t'_1; \ id'_2 : t'_2; \ \ldots; \ id'_p : t'_p) \to t')$, then

$$\tilde{\tilde{s}} = Signat((id'_1 \ isr_1 \ \tilde{\tilde{t}}'_1; \ id'_2 \ isr_2 \ \tilde{\tilde{t}}'_2; \ \ldots; \ id'_p \ isr_p \ \tilde{\tilde{t}}'_p) \to \tilde{\tilde{t}}')$$

Hereby, $\tilde{\tilde{t}}'$ specifies the generalized type of the returned value(s) of the operation and $\tilde{\tilde{t}}'_i$, $i = 1, 2, \ldots, p$ specify the generalized types of the input parameters. The variable copula *isr* then denotes the generalized constraint, which constrains the valid return values (of the generalized type $\tilde{\tilde{t}}'$) and defines their interpretation. The variable copula isr_i, $i = 1, 2, \ldots, p$ denote the generalized constraints, which constrain and interpret the parameter values of the operation. The same kinds of generalized constraints *ise*, *is* and *isv* have been considered.

Each type specification $\tilde{\tilde{is}} \in \tilde{\tilde{T}}$ is formally defined as specified in subsection 4.3.1. The syntax rules are defined as follows:

Definition 6 (Syntax rules – generalized) *Let ID denote the set of valid identifiers. The set $\tilde{\tilde{T}}$ of expressions that satisfy the syntax of a generalized type is inductively defined as:*

- *Generalized basic types.*

  - $\{\tilde{\tilde{Integer}}, \tilde{\tilde{Real}}, \tilde{\tilde{Boolean}}, \tilde{\tilde{Octet}}, \tilde{\tilde{String}}, \tilde{\tilde{Any}}\} \subset \tilde{\tilde{T}}$.

  - *If $id \in ID$ and $\{id_1, id_2, \ldots, id_n\} \subseteq ID$, then*
    $$\tilde{\tilde{Enum}} \ id(id_1, id_2, \ldots, id_n) \in \tilde{\tilde{T}}$$
    (generalized enumeration type).

    The identifier $id \in ID$ identifies the generalized enumeration type, *whereas* $(id_1, id_2, \ldots, id_n)$ *represents the ordered sequence of identifiers, which is described by the type.*

- *Generalized constructed types.*

  - *If* $\tilde{\tilde{t}} \in \tilde{\tilde{T}} \setminus \{\tilde{A}ny\}$, *then* $Co\tilde{\tilde{l}}lect(\tilde{\tilde{t}}) \in \tilde{\tilde{T}}$ *(generalized collection type).*

  - *If* $id \in ID$, $\{id_1, id_2, \ldots, id_n\} \subseteq ID$ *and* $\tilde{\tilde{t}}_i \in \tilde{\tilde{T}} \setminus \{\tilde{A}ny\}$, $1 \leq i \leq n$, *then* $St\tilde{\tilde{r}}uct\ id(id_1\ isr_1\ \tilde{\tilde{t}}_1;\ id_2\ isr_2\ \tilde{\tilde{t}}_2;\ \ldots;\ id_n\ isr_n\ \tilde{\tilde{t}}_n) \in \tilde{\tilde{T}}$
  (generalized tuple-type).

    The identifier id identifies the generalized tuple-type, whereas $(id_1\ isr_1\ \tilde{\tilde{t}}_1;\ id_2\ isr_2\ \tilde{\tilde{t}}_2;\ \ldots;\ id_n\ isr_n\ \tilde{\tilde{t}}_n)$ *represents the components of the generalized tuple-type. Each component* $id_i\ isr_i\ \tilde{\tilde{t}}_i$, $1 \leq i \leq n$ *is characterized by an identifier* id_i *a variable copula* isr_i, *which identifies a generalized constraint and a generalized type* $\tilde{\tilde{t}}_i$.

  - *Let* $\tilde{\tilde{V}}_{signat}$ *denote the set of all valid generalized operation signatures, which is defined as follows:*
    * *If* $\tilde{\tilde{t}}' \in \tilde{\tilde{T}}$ *then* $Signat((\ ) \to \tilde{\tilde{t}}') \in \tilde{\tilde{V}}_{signat}$.
    * *If* $\tilde{\tilde{t}}' \in \tilde{\tilde{T}}$, $\{id_1', id_2', \ldots, id_p'\} \subseteq ID$ *and* $\tilde{\tilde{t}}_i' \in \tilde{\tilde{T}}$, $1 \leq i \leq p$, *then*
    $$Signat((id_1'\ isr_1\ \tilde{\tilde{t}}_1';\ id_2'\ isr_2\ \tilde{\tilde{t}}_2';\ \ldots;\ id_p'\ isr_p\ \tilde{\tilde{t}}_p') \to \tilde{\tilde{t}}') \in$$
    $\tilde{\tilde{V}}_{signat}$.

    Hereby, id_i', isr_i *and* $\tilde{\tilde{t}}_i'$, $1 \leq i \leq p$ *are respectively the names, the variable copula which identify a generalized constraint and the expected generalized types of the input parameters of the operation and* $\tilde{\tilde{t}}'$ *is the generalized type of the returned value(s) of the operation.*

    If $id \in ID$, $\{\widehat{id}_1, \widehat{id}_2, \ldots, \widehat{id}_m\} \subset ID \setminus \{id\}$, $\{id_1, id_2, \ldots, id_n\} \subset ID$ *and* $\tilde{\tilde{s}}_i \in \tilde{\tilde{T}} \cup \tilde{\tilde{V}}_{signat} \setminus \{\tilde{A}ny\}$, $1 \leq i \leq n$, *then*

    * $C\tilde{\tilde{l}}ass\ id(id_1\ isr_1\ \tilde{\tilde{s}}_1;\ id_2\ isr_2\ \tilde{\tilde{s}}_2;\ \ldots;\ id_n\ isr_n\ \tilde{\tilde{s}}_n) \in \tilde{\tilde{T}}$

    * $C\tilde{\tilde{l}}ass\ id : \widehat{id}_1, \widehat{id}_2, \ldots, \widehat{id}_m(\ ) \in \tilde{\tilde{T}}$

    * $Class\ id : \widehat{id}_1, \widehat{id}_2, \ldots, \widehat{id}_m(id_1\ isr_1\ \tilde{\tilde{s}}_1;\ id_2\ isr_2\ \tilde{\tilde{s}}_2;\ \ldots;\ id_n\ isr_n\ \tilde{\tilde{s}}_n)$
    $$\in \tilde{\tilde{T}}$$

      (generalized object type)

    The identifier id identifies the generalized object type, the identifiers $\widehat{id}_i$, $1 \leq i \leq m$ *are the identifiers of the supertypes of the generalized object type, whereas* $(id_1 \ isr_1 \ \tilde{\tilde{s}}_1; \ id_2 \ isr_2 \ \tilde{\tilde{s}}_2; \ \dots; \ id_n \ isr_n \ \tilde{\tilde{s}}_n)$ *represents the characteristics of the generalized object type. (Beside these characteristics, the generalized object type can inherit other characteristics from its parent types.)*

□

The full semantics of the generalized types $\tilde{\tilde{ts}} \in \tilde{\tilde{T}}$ (cf. definition 6) are defined by providing an appropriate definition for the set of domains $D_{\tilde{\tilde{ts}}}$, the domain $dom_{\tilde{\tilde{ts}}}$ of the generalized type, the set of operators $O_{\tilde{\tilde{ts}}}$ and the set of axioms $A_{\tilde{\tilde{ts}}}$. Full detailed definitions are beyond the scope of this chapter. Instead an informal description is given:

- *Generalized basic types.* The definition of the generalized basic types is as follows:
    - The domain of the $\tilde{Integer}$-type is defined as

    $$dom_{\tilde{Integer}} = \tilde{\wp}(\tilde{\wp}(dom_{Integer}))$$

    The set of operators $O_{\tilde{Integer}}$ consists of the operators $\tilde{\tilde{=}}, \tilde{\tilde{\neq}}, \tilde{\tilde{<}}, \tilde{\tilde{>}}, \tilde{\tilde{\leq}}, \tilde{\tilde{\geq}}, \tilde{\tilde{+}}, \tilde{\tilde{-}}, \tilde{\tilde{*}}, \tilde{div}$ and $\tilde{mod}$, (which all have been defined as generalized operators), a set of specific operators $O^{specific}$ and the bottom operation $\bot$, which always results in an "undefined" generalized integer value (cf. example 8).

    - The domain of the $\tilde{Real}$-type is defined as

    $$dom_{\tilde{Real}} = \tilde{\wp}(\tilde{\wp}(dom_{Real}))$$

    whereas the set of operators $O_{\tilde{Real}}$ consists of the operators $\tilde{\tilde{=}}, \tilde{\tilde{\neq}}, \tilde{\tilde{<}}, \tilde{\tilde{>}}, \tilde{\tilde{\leq}}, \tilde{\tilde{\geq}}, \tilde{\tilde{+}}, \tilde{\tilde{-}}, \tilde{\tilde{*}}$ and $\tilde{/}$ (which all have been defined as generalized operators), a set of specific operators $O^{specific}$ and the bottom operation $\bot$, which always returns an "undefined" generalized real value.

    - The domain of the $\tilde{Boolean}$-type is defined as

    $$dom_{\tilde{Boolean}} = \tilde{\wp}(\{\{(x,1)\} \mid x \in dom_{Boolean}\})$$

    since vague or imprecise Boolean values have been considered as not being meaningful. The set $O_{\tilde{Boolean}}$ consists of the operators $\tilde{\tilde{=}}, \tilde{\tilde{\neq}}, \tilde{\tilde{\wedge}}, \tilde{\vee}$ and $\tilde{\tilde{\neg}}$ (which all have been defined as generalized operators), a set of specific operators $O^{specific}$ and the bottom operation $\bot$, which always returns an "undefined" generalized Boolean value.

- The domain of the $\tilde{O}ctet$-type is defined as

$$dom_{\tilde{O}ctet} = \tilde{\wp}(\{\{(x,1)\} \mid x \in dom_{Octet}\})$$

since vague or imprecise bytes have been considered as not being meaningful (uncertainty about the value of a byte is still allowed). The set $O_{\tilde{O}ctet}$ consists of the operators $\tilde{\approx}$, $\tilde{\neq}$, $\tilde{<}$, $\tilde{>}$, $\tilde{\leq}$, $\tilde{\geq}$, $\tilde{\wedge}$, $\tilde{\vee}$, $\tilde{\approx}$, $Shift\_left$ and $Shift\_right$ (which all have been defined as generalized operators), a set of specific operators $O^{specific}$ and the bottom operation $\bot$, which always returns an "undefined" generalized byte.

- The domain of the $\tilde{String}$-type is defined as

$$dom_{\tilde{String}} = \tilde{\wp}(\tilde{\wp}(dom_{String}))$$

The set $O_{\tilde{String}}$ consists of the operator $\tilde{substr}$ (which has been defined as an extended operator), the operators $\tilde{\approx}$, $\tilde{\neq}$, $\tilde{<}$, $\tilde{>}$, $\tilde{\leq}$, $\tilde{\geq}$, $\tilde{concat}$ and $\tilde{sizeof}$ (which all have been defined as generalized operators), a set of specific operators $O^{specific}$ and the bottom operation $\bot$, which always results in an "undefined" generalized string value.

- The domain of a generalized enumeration type with specification $\tilde{Enum}\ id(id_1, id_2, \ldots, id_n)$ is defined as

$$dom_{\tilde{Enum}\ id(id_1,id_2,\ldots,id_n)} = \tilde{\wp}(\tilde{\wp}(dom_{Enum\ id(id_1,id_2,\ldots,id_n)}))$$

The set $O_{\tilde{Enum}\ id(id_1,id_2,\ldots,id_n)}$ consists of the operators $\tilde{\approx}$, $\tilde{\neq}$, $\tilde{<}$, $\tilde{>}$, $\tilde{\leq}$ and $\tilde{\geq}$ (which all have been defined as generalized operators), a set of specific operators $O^{specific}$ and the bottom operation $\bot$, which always returns an "undefined" domain value of the generalized enumeration type.

- The domain of the $\tilde{Any}$-type is defined as the singleton

$$\{(\{\{(\bot_{Any},1)\}, 1)\}$$

Its corresponding set of operators $O_{\tilde{Any}}$ is the singleton

$$\{\bot : \to dom_{\tilde{Any}}\}$$

• *Generalized constructed types.* Generalized constructed types are characterized by composed generalized domain values.

- *Generalized collection type.* The generalized collection type is a generic type, indicated by the type generator $\widetilde{Collect}$ and a generalized type parameter $\widetilde{\tilde{t}}$. The domain of a generalized collection type $\widetilde{Collect}(\widetilde{\tilde{t}})$ is defined as

$$dom_{\widetilde{Collect}(\widetilde{\tilde{t}})} = \tilde{\wp}(\tilde{\wp}(dom_{\widetilde{Collect}(\widetilde{\tilde{t}})}))$$

  where $dom_{\widetilde{Collect}(\widetilde{\tilde{t}})}$ consists of the "undefined" domain value $\perp_{\widetilde{Collect}(\widetilde{\tilde{t}})}$ and of sets of elements of the domain of generalized type $\widetilde{\tilde{t}}$. The associated set $O_{\widetilde{Collect}(\widetilde{\tilde{t}})}$ consists of the operators $\widetilde{=}$, $\widetilde{\neq}$, $\widetilde{\subset}$, $\widetilde{\supset}$, $\widetilde{\subseteq}$, $\widetilde{\supseteq}$, *cardinality*, *is_empty*, $\widetilde{\cup}$, $\widetilde{\cap}$, $\widetilde{\setminus}$ and *contains_element* (which all have been defined as generalized operators), a set of specific operators $O^{specific}$ and the bottom operation $\perp$, which always returns an "undefined" domain value of the generalized collection type.

- *Generalized tuple-type.* A generalized tuple-type is structured and consists of a fixed number of components id_i isr_i $\widetilde{ts}_i$, $i = 1, 2, \ldots, n$. The domain of a generalized tuple-type

$$\widetilde{Struct}\ id(id_1\ isr_1\ \widetilde{ts}_1;\ id_2\ isr_2\ \widetilde{ts}_2;\ \ldots;\ id_n\ isr_n\ \widetilde{ts}_n) \in \widetilde{\tilde{T}}$$

  is defined as

$$dom_{\widetilde{Struct}\ id(id_1\ isr_1\ \widetilde{ts}_1;\ id_2\ isr_2\ \widetilde{ts}_2;\ \ldots;\ id_n\ isr_n\ \widetilde{ts}_n)} = $$
$$\tilde{\wp}(\tilde{\wp}(dom_{\widetilde{Struct}\ id(id_1\ isr_1\ \widetilde{ts}_1;\ id_2\ isr_2\ \widetilde{ts}_2;\ \ldots;\ id_n\ isr_n\ \widetilde{ts}_n)}))$$

  where $dom_{\widetilde{Struct}\ id(id_1\ isr_1\ \widetilde{ts}_1;\ id_2\ isr_2\ \widetilde{ts}_2;\ \ldots;\ id_n\ isr_n\ \widetilde{ts}_n)}$ contains the "undefined" domain value $\perp_{\widetilde{Struct}\ id(id_1\ isr_1\ \widetilde{ts}_1;\ id_2\ isr_2\ \widetilde{ts}_2;\ \ldots;\ id_n\ isr_n\ \widetilde{ts}_n)}$ and composed values, which each consist of n values $id_i : v_i$, with $v_i \in dom_{\widetilde{ts}_i}$, $i = 1, 2, \ldots, n$ thereby satisfying the generalized constraint denoted by the associated copula isr_i. The set of operators

$$O_{\widetilde{Struct}\ id(id_1\ isr_1\ \widetilde{ts}_1;\ id_2\ isr_2\ \widetilde{ts}_2;\ \ldots;\ id_n\ isr_n\ \widetilde{ts}_n)}$$

  consists of the operators $\widetilde{=}$ and $\widetilde{\neq}$ (which both have been defined as generalized operators), a set of specific operators $O^{specific}$, the operators *set_component* and *get_component* and the bottom operation $\perp$, which always returns an "undefined" domain value of the generalized tuple-type.

- *Generalized object type.* Each generalized object type is characterized by a fixed number of characteristics id_i isr_i $\widetilde{\tilde{s}}_i$, $i = 1, 2, \ldots, n$. Some of them can be inherited from the parent types of the generalized object type. The inheritance-based type-subtype relationships between generalized object types have been defined by means of a partial ordering function $\tilde{\prec}$, which is defined over the set $\tilde{T}$ of generalized type specifications. ($\widehat{id} \; \tilde{\prec} \; id$ denotes that "generalized object type id inherits all characteristics of generalized object type $\widehat{id}$".)

The domain of a generalized object type with specification $\tilde{ts}$ is defined as

$$dom_{\tilde{ts}} = \tilde{\wp}(\tilde{\wp}(dom_{ts}))$$

and thus consists of a set of level-2 fuzzy sets, which are defined over a set with the "undefined" domain value $\perp_{ts}$, references to generalized objects of the generalized type $\tilde{ts}$ and references to generalized objects of the subtypes of $\tilde{ts}$. Each generalized object is composed and contains a value $id : v$, for each of the (inherited) attributes $id \; isr \; \tilde{s}$, $\tilde{s} \in \tilde{T}$ of its type. Hereby, $v \in dom_{\tilde{s}}$. Furthermore, v satisfies the generalized constraint denoted by the associated copula isr. The set of operators associated with a given generalized object type is the union of a set of implicitly defined operators and a set of explicitly defined operators. The set of implicitly defined operators consists of the operators $\tilde{=}$ and $\tilde{\neq}$ (which both have been defined as generalized operators), a set of specific operators $O^{specific}$, the operators $\widetilde{set\_attribute}$ and $\widetilde{get\_attribute}$ and the bottom operation $\perp$, which always returns an "undefined" domain value of the generalized object type. The explicitly defined operators are the (inherited) methods $id \; isr \; \tilde{s}$, $\tilde{s} \in \tilde{V}_{signat}$ of the generalized object type.

The generalized type system GTS, which defines all the valid generalized types supported by the presented generalized data model, is defined as:

Definition 7 (Generalized type system) *The generalized type system GTS is formally defined by the quintuple*

$$GTS = [ID, \tilde{T}, P, \tilde{f}_{impl}, \tilde{\prec}]$$

where

- *ID is the set of the valid identifiers,*
- *$\tilde{T}$ is the set of valid specifications of generalized types (cf. definition 6),*
- *P is the infinite set of all type implementations (cf. definition 5),*
- *$\tilde{f}_{impl}$ is the mapping function, which maps each generalized type specification onto its corresponding set of type implementations (cf. definition 5) and*

- $\tilde{\tilde{\prec}} \;:\; \tilde{\tilde{T}}^2 \to \{\,True,\,False\,\}$ *is the partial ordering which is used to define the inheritance-based type-subtype relationships between generalized object types.*

$\square$

The examples 3 and 4 of subsection 4.2.2 can be rebuilt using the presented generalized type system:

Example 10. The generalized type system allows for the following definitions, which are intended to describe a (simplified) generalized type for employees. With the generalized tuple-types with specification

$$\tilde{\tilde{Struct}}\ GTAddress(Street\ ise\ \tilde{\tilde{String}};\ City\ ise\ \tilde{\tilde{String}};\ Zip\ ise\ \tilde{\tilde{Integer}}),$$

$$\tilde{\tilde{Struct}}\ GTCompany(Name\ ise\ \tilde{\tilde{String}};\ Location\ isv\ \tilde{\tilde{String}})$$

and

$$\tilde{\tilde{Struct}}\ GTWorks(Company\ ise\ GTCompany;\ Percentage\ is\ \tilde{\tilde{Real}})$$

and the generalized enumeration type with specification

$$\tilde{\tilde{Enum}}\ GTLang(Dutch, French, English, German, Spanish, Italian)$$

the specifications of the generalized object types

$$[GTPerson, \tilde{\tilde{f}}_{impl}(GTPerson)] \text{ and } [GTEmployee, \tilde{\tilde{f}}_{impl}(GTEmployee)]$$

can be defined as:

$$\tilde{\tilde{Class}}\ GTPerson($$

$$\quad Name\ ise\ \tilde{\tilde{String}};$$

$$\quad Age\ is\ \tilde{\tilde{Integer}};$$

$$\quad Address\ ise\ GTAddress;$$

$$\quad Languages\ isv\ GTLang;$$

$$\quad Children\ ise\ \tilde{\tilde{Collect}}(GTPerson);$$

$$\quad Add\_child\ ise\ Signat\ ((New\_child\ ise\ GTPerson) \to \tilde{\tilde{Any}}))$$

and

$$\tilde{\tilde{Class}}\ GTEmployee : GTPerson($$

$$\quad EmployeeID\ ise\ \tilde{\tilde{String}};$$

$$\quad Works\_for\ ise\ \tilde{\tilde{Collect}}(GTWorks))$$

$\diamond$

Example 11. The instances of the generalized object type *GTPerson* of example 10 are level-2 fuzzy sets which consist of references to "GTPerson"-objects and of references to "GTEmployee"-objects (since *GTEmployee* is a subtype of *GTPerson*). The instances of the generalized object type *GTEmployee* are level-2 fuzzy sets of references to "GTEmployee"-objects.

Examples of "GTPerson"-objects are:

$$(Name : \{(\{(\text{``}Ann\text{''}, 1)\}, 1)\};$$
$$Age : \{(\{(12, 0.4), (13, 0.8), (14, 1), (15, 0.6)\}, 1)\};$$
$$Address : \{(\{((Street : \{(\{(\text{``}ChurchStreet, 12\text{''}, 1)\}, 1)\};$$
$$City : \{(\{(\text{``}Brussels\text{''}, 1)\}, 1)\};$$
$$Zip : \{(\{(1000, 1)\}, 1)\}), 1)\}, 1)\};$$
$$Languages : \{(\{(Dutch, 1), (French, 0.3)\}, 1)\};$$
$$Children : \{(\{((), 1)\}, 1)\})$$

Interpretation: *The person named Ann is about 14 years old, lives in the Church Street 12, 1000 Brussels and has a knowledge of both the Dutch and French languages. Her Dutch knowledge is fluent; her French knowledge is rather premature. Ann has no children*

and

$$(Name : \{(\{(\text{``}Joe\text{''}, 1)\}, 1)\};$$
$$Age : \{(\{(18, 1)\}, 1), (\{(21, 1)\}, 0.6)\};$$
$$Address : \{(\{((Street : \{(\{(\text{``}ChurchStreet, 12\text{''}, 1)\}, 1)\};$$
$$City : \{(\{(\text{``}Brussels\text{''}, 1)\}, 1)\};$$
$$Zip : \{(\{(1000, 1)\}, 1)\}), 1)\}, 1)\};$$
$$Languages : \{(\{(Dutch, 1), (French, 0.6), (English, 0.6)\}, 1)\};$$
$$Children : \{(\{((), 1)\}, 1)\})$$

Interpretation: *The person named Joe is possibly 18 years old or less possibly 21 years old. Joe lives in the Church Street 12, 1000 Brussels and has a knowledge of the Dutch, French and English languages. His Dutch knowledge is fluent; his French and English knowledge is at school level. Joe has no children*

An example of a "GTEmployee"-object is:

$$(Name : \{(\{(\text{``}John\text{''}, 1)\}, 1)\};$$
$$Age : \{(\{(42, 1)\}, 1)\};$$
$$Address : \{(\{((Street : \{(\{(\text{``}ChurchStreet, 12\text{''}, 1)\}, 1)\};$$

$$City : \{(\{(\text{``Brussels''}, 1)\}, 1)\};$$
$$Zip : \{(\{(\{(1000, 1)\}, 1)\}), 1)\}, 1)\};$$
$$Languages : \{(\{(Dutch, 1), (French, 0.8), (English, 0.8)\}, 1),$$
$$(\{(Dutch, 1), (French, 0.8), (English, 0.8), (Italian, 0.3)\}, 1)\};$$
$$Children : \{(\{((\{(\{(ref\_to\_Ann, 1)\}, 1)\}, \{(\{(ref\_to\_Joe, 1)\}, 1)\}), 1)\}, 1)\};$$
$$EmployeeID : \{(\{(\text{``ID254''}, 1)\}, 1)\};$$
$$Works\_for : \{(\{((\{(\{((Company : \{(\{((Name : \{(\{(\text{``IBM''}, 1)\}, 1)\};$$
$$Location : \{(\{(\text{``Brussels''}, 1)\}, 1)\}), 1)\}, 1)\};$$
$$Percentage : \{(\{(\{(100, 1)\}, 1)\}), 1)\}, 1)\}), 1)\}, 1)\})$$

Interpretation: *The employee named John is 42 years old. He lives in the Church Street 12, 1000 Brussels and has a fluent knowledge of the Dutch language. John also has a good knowledge of French and English. Possibly he has a premature knowledge of Italian too. John has two children: Ann (with reference ref_to_Ann) and Joe (with reference ref_to_Joe). His employee identification number is ID254. Joe has a full time job at IBM in Brussels.*

As it is illustrated with the "GTEmployee"-object, the nesting of level-2 fuzzy sets can result in complex notations. A shorter, more readable notation is obtained by replacing the level-2 fuzzy sets $\{(\{(v, 1)\}, 1)\}$ with the shorthand v. The "GTEmployee"-object then becomes

$(Name : \text{``John''};$

$\quad Age : 42;$

$\quad Address : (Street : \text{``ChurchStreet''}, 12''; \ City : \text{``Brussels''}; \ Zip : 1000);$

$\quad Languages : \{(\{(Dutch, 1), (French, 0.8), (English, 0.8)\}, 1),$

$\qquad\qquad\qquad (\{(Dutch, 1), (French, 0.8), (English, 0.8), (Italian, 0.3)\}, 1)\};$

$\quad Children : (ref\_to\_Ann, ref\_to\_Joe);$

$\quad EmployeeID : \text{``ID254''};$

$\quad Works\_for : ((Company : (Name : \text{``IBM''}; \ Location : \text{``Brussels''});$

$\qquad\qquad\qquad Percentage : 100)))$

◇

4.4   Conclusions

A formal framework for the definition of a generalized object-oriented data model for the modeling of both crisply and non-crisply described information has been presented. This framework is obtained as a generalization of an ordinary type system, which is consistent with the specifications of the object model presented by the Object Management Group (OMG). Generalizing a type system, so that it encompasses an ordinary type system, allows to handle information in a more "natural" and transparant way: each generalized type, supported by the generalized type system, is suited to handle both crisply described domain values and non-crisply described domain values (i.e. uncertain, imprecise, vague and incomplete domain values, and any valid combination of these) in a uniform way.

The definitions of the types supported by the ordinary type system are related to domains, sets of domains and sets of operators. On the basis of the generalization of a type is the representation of the domain values by level-2 fuzzy sets, which are all defined over the ordinary domain. Level-2 fuzzy sets allow to model uncertainty and fuzziness in two distinct nested layers: the "inner layer" corresponding to the fuzzy description of the data themselves, the "outer layer" to the (un)certainty about the existence of the data. This allows to represent "crisp" domain values, uncertainty, imprecision or vagueness about these values, and any valid combination of these in a uniform and advantageous way.

The use of level-2 fuzzy sets as domain values has an impact on the (definition of) the operators of a generalized type. Three kinds of operators are distinguished: extended operators, which have the same behavior as their crisp counterpart for "crisp" arguments and result in an "undefined" domain value if at least one of the arguments represents a non-crisp value; generalized operators, which are obtained as a generalization of their crisp counterpart by applying Zadeh's extension principle two successive times; and specific operators, which have been provided for the handling of level-2 fuzzy sets and their "inner-level" fuzzy sets.

The presented generalized type system supports the definitions of the generalized counterparts of each of the ordinary types supported by the OMG-consistent type system. It can be applied within the context of (the formal definition of) any data model for software and data engineering applications, which additionally need to cope with both fuzziness and uncertainty.

References

1. J.F. Baldwin and T.P. Martin. Fuzzy classes in object-oriented logic programming. In *Proceedings of the Fifth IEEE International Conference on Fuzzy Systems*, pages 1358–1364, 1996.
2. M. Black. Vagueness: an exercise in logical analysis. *Philosphy of Science*, 4(4):427–455, 1937.
3. T.H. Cao and P.N. Creasy. Fuzzy types: a framework for handling uncertainty about types of objects. *International Journal of Approximate Reasoning*, 25:217–253, 2000.
4. V. Cross. Towards a unifying framework for a fuzzy object model. In *Proceedings of the Sixth IEEE International Conference on Fuzzy Systems*, pages 85–92, 1997.

5. G. de Tré and R. de Caluwe. The application of generalized constraints to object-oriented database models. *Mathware & Soft Computing*, VII:2–3, 245–255, 2000.
6. G. de Tré, R. de Caluwe, and B. Van der Cruyssen. A generalised object-oriented database model. In G. Bordogna and G. Pasi, editors, *Recent Issues on Fuzzy Databases: Studies in Fuzziness and Soft Computing*, pages 155–182. Physica-Verlag, Heidelberg, Germany, 2000.
7. D. Dubois and H. Prade. *Possibility Theory*. Plenum Press, New York, USA, 1988.
8. D. Dubois and H. Prade. The tree semantics of fuzzy sets. *Fuzzy Sets and Systems*, 90(2):141–150, 1997.
9. D. Dubois, H. Prade, and J.P. Rossazza. Vagueness, typicality, and uncertainty in class hierarchies. *International Journal of Intelligent Systems*, 6:167–183, 1991.
10. J. Early. Toward an understanding of data structures. *Communications of the ACM*, 14(10):617–627, 1971.
11. S. Gottwald. Set theory for fuzzy sets of higher level. *Fuzzy Sets and Systems*, 2(2):125–151, 1979.
12. M. Guttman and J.R. Matthews. *The Object Technology Revolution (Now is the Time to Lead, Follow, or Get Out of the Way)*. John Wiley & Sons Inc. and Object Management Group, New York, New York, USA, 1995.
13. G.J. Klir and B. Yuan. *Fuzzy Sets and Fuzzy Logic: Theory and Applications*. Prentice Hall Inc., New Jersey, USA, 1995.
14. T. Korson and J.D. McGregor. Understanding object-oriented: A unifying paradigm. *Communications of the ACM*, 13(9):40–60, 1990.
15. H.F. Korth and A. Silberschatz. Database research faces the information explosion. *Communications of the ACM*, 40(2):139–142, 1997.
16. G. Lausen and G. Vossen. *Models and Languages of Object-Oriented Databases*. International Computer Science Series. Addison-Wesley Publishing Company, Harlow, England, 1998.
17. A. Motro. Management of uncertainty in database systems. In W. Kim, editor, *Modern Database Systems: The object model, interoperability and beyond*. Addison-Wesley Publishing Company, Reading, Massachusetts, USA, 1995.
18. S. Parsons. Current approaches to handling imperfect information in data and knowledge bases. *IEEE Transactions on Knowledge and Data Engineering*, 8(3):353–372, 1996.
19. N. Rescher. *Many-Valued Logic*. McGraw-Hill, New York, USA, 1969.
20. J. Siegel. *CORBA: Fundamentals and Programming*. John Wiley & Sons Inc. New York, New York, USA, 1996.
21. L.A. Zadeh. Fuzzy sets. *Information and Control*, 8(3):338–353, 1965.
22. L.A. Zadeh. The concept of a linguistic variable and its application to approximate reasoning (parts I, II and III). *Information Sciences*, 8:199–251, 301–357 and 9:43–80, 1975.
23. L.A. Zadeh. Fuzzy logic and approximate reasoning. *Synthese*, 30(1):407–428, 1975.
24. L.A. Zadeh. Fuzzy sets as a basis for a theory of possibility. *Fuzzy Sets and Systems*, 1(1):3–28, 1978.
25. L.A. Zadeh. Fuzzy logic = computing with words. *IEEE Transactions on Fuzzy Systems*, 4(2):103–111, 1996.

26. L.A. Zadeh. Toward a theory of fuzzy information granulation and its centrality in human reasoning and fuzzy logic. *Fuzzy Sets and Systems*, 90:111–127, 1997.
27. L.A. Zadeh. A new direction in fuzzy logic – toward a computational theory of perceptions. In R.N. Davé and T. Sudkamp, editors, *Proceedings of the 18th NAFIPS International Conference*, pages 1–4. IEEE, New York, New York, USA, June 10-12 1999.

5 Modelling Imperfect Spatial Information in a Fuzzy Object Oriented Database

Gloria Bordogna[1] and Sergio Chiesa[2]

[1] CNR-ITIM via Ampére 56, 20133 - Milano, Italy
email:gloria@itim.mi.cnr.it
[2] CNR-IDPA ex CSGAQ sezione di Bergamo, p.zza Cittadella 4 - Bergamo, Italy

5.1 Introduction

The adoption of an object oriented database model has been proposed as appropriate for GIS applications by several researchers [1,5,12]: it allows to develop a set of data types that maintain a one-to-one correspondence with geographic entities thus directly reflecting the human perception of the reality.

On the other side, imperfection is recognised as being an endemic feature of geographic information [5] and the ability to store and manage large collections of imperfect spatial information is a necessity for many applications and for many categories of users: for example, land value assessors have to represent empirical entities with undefined boundaries in a well-bounded way; pilots and fighters need objects with sharp boundaries while well-bounded entities cannot be adequately observed and represented; planners and resource managers have entities that neither have, nor can easily be assigned well-defined boundaries, while they need well-defined boundaries [11].

The automatic management of imperfect spatial information requires new capabilities for current Geographic Information Systems (GIS) [1]. In facts, the two conceptual geographic data models for representing spatial information used by all current GISs, the exact object based model and the field based model, force to an extreme abstraction of reality [5]. In the exact object based model, individual objects representation with well determined boundaries and attributes is stressed. In the second model the space is represented as a continuous field of data values. These two dual models present some shortcomings in many real situations:

- they do not allow to model spatial changes;
- they do not allow to represent the imprecision or uncertainty neither in the attribute values nor in the description and identification of real spatial entities.

In this paper, after proposing a representation of imperfect spatial information consistent with possibility theory and fuzzy set theory we suggest an approach to represent and manage such information in a fuzzy object oriented database. The concept of linguistic variable and the theory of possibility

[28,13] have been successfully applied to manage vague and uncertain information in a unique formal context in fuzzy databases [4,20,6,24,2]: by adopting this framework, vague and uncertain attribute values can be specified as fuzzy subsets of the attribute domain. These fuzzy subsets are interpreted as possibility distributions so that the degree of membership of each domain value expresses its degree of possibility to be the actual manifestation. Fuzzy pattern matching and/or similarity based matching are then adopted to define the query evaluation mechanism [4,20,14]. The problem of representing imperfect spatial information in databases has have been faced by several authors [5,12,26,15,22,9,21,27]: to this aim, some approaches proposed the use of fuzzy databases [9,21,27]. The novelty of our approach is the adoption of linguistic qualifiers, defined based on the concept of linguistic variable, to describe properties of spatial entities [29]. Linguistic qualifiers are a habit in many disciplines and the ability to manage them directly can greatly improve the expressive power of the representation. Further some guidelines for defining a flexible query language for imperfect spatial information are discussed. In particular, vague topological constraints between fuzzy and indeterminate spatial entities are introduced.

In this paper, in section 5.2. the management of imperfect information in fuzzy databases is briefly summarized. Then in section 5.3. the sources and characteristics of imperfection of spatial information are analyzed and their representation within possibility theory proposed. In section 5.4 new data types for representing imperfect spatial information are defined in the context of a fuzzy object oriented data model. In section 5.5, the introduction of the problem of flexible querying of fuzzy spatial information is described and an evaluation mechanism of topological constraints against imperfect spatial information is proposed. Finally in the conclusions the main results are summarized.

5.2   Imperfect information in fuzzy databases

Fuzzy database models have been defined for dealing with imperfect information, either in the data, or in the queries, and in both data and queries [4,20,6,24,2]. Imperfect data are represented in a fuzzy database differently according to the kind of imperfection.

Vagueness stems by two distinct situations: the first one is the impossibility to make a sharp distinction regarding the presence or absence of a given property that is gradual by its nature. The specification of the limit between full presence and full absence of the property can then be defined only on conventional basis. To represent gradual properties in a fuzzy database strengthened relationships between objects have been defined as fuzzy relations in which the degree in [0,1], or alternatively the linguistic degree defined on an ordinal scale such as {*none, low, medium, high, full*}, expresses the strength of association between the two objects [2]: for example to represent the strength of knowledge of a language by a person one can define the fuzzy relation *Known languages* so as to associate with each *person* in the database (ex. *Mary*) a value of the attribute *languages* (ex. *English*) by specifying a degree (ex. *high*). The second source of vagueness in data is either vague and incomplete knowledge or the need to synthesize considerable knowledge about the reality to be modelled. This vagueness leads to the impossibility or inability to select a precise value of an attribute domain as attribute value. In these situations an imprecise or vague attribute value can be specified with a higher granularity with respect to the crisp values. *Imprecision* occurs in the case that subsets of the attribute domain are specified as attribute value. Often vague attribute values are expressed by linguistic values (ex. *young,*) the semantics of which is defined by fuzzy subsets of the attribute domain [4].

Uncertainty is related to the doubt about the conformity of the stored information with the reality that is described. It may derive by either the lack of reliability of the sources of knowledge or the incompleteness of the available knowledge. It can be represented by specifying a confidence degree $\varepsilon \in [0,1]$ associated with a precise or vague attribute value.

The concept of linguistic variable and the theory of possibility have been applied to manage imprecise/vague and uncertain information in a unique formal context [28,13]. By adopting this framework in a fuzzy database, uncertain and vague/imprecise attribute values can be specified as fuzzy subsets of the attribute domain; these fuzzy subsets are interpreted as possibility distributions; in this case the degree of membership of each domain value expresses its degree of possibility to be the actual manifestation. In Fig. 5.1 the possibility distribution μ_{high} representing the vague value *"high"* of the attribute *depth* is shown.

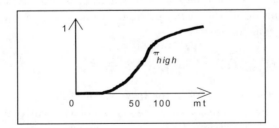

Fig. 5.1. Representation of the value *high* of the linguistic variable *depth*

Fuzzy pattern matching constitutes a general framework in which to define the query evaluation mechanism in a fuzzy database [4,20,6,24,2,14]. From the possibility distributions $\pi_{A(x)}(.)$ representing the semantics of the vague attribute value A(x), and the fuzzy subset $\mu_Q(.)$ representing a query constraint Q, it is possible to compute the two measures $\Pi(Q, A(x))$ and

$N(Q, A(x))$ expressing the possibility and the necessity by which the value $A(x)$ satisfies the condition Q:

$$\mu_{\Pi_Q}(x) = \Pi(Q;\ A(x)) = Sup_{i \in U}\ min(\mu_Q(i), \pi_{A(x)}(i))$$

$$\mu_{N_Q}(x) = N(Q;\ A(x)) = Inf_{i \in U \cup \{e\}}\ max(\mu_Q(i), 1 - \pi_{A(x)}(i))$$

in which D is the basic domain of the A attribute and $\{e\}$ represents the case when A does not apply to x. $\Pi(Q, A(x))$ estimates to what extent there is a value restricted by $\pi_{A(x)}(.)$ compatible with Q and $N(Q, A(x))$ to what extent all the values more or less possible for $A(x)$ are included in Q.

There are other approaches to evaluate fuzzy queries against imperfect data in fuzzy databases, mainly named similarity based models [20]. In the ordinal approaches [20], the similarity between the vague query constraint Q and the ill-known attribute value A is evaluated based on fuzzy relations (similarity or proximity relations) defined in the form of matrices. In the numeric approaches [24,3], given a query condition Q and a vague attribute value $A(x)$ one can obtain a degree of satisfaction of the query Q by x by computing the measure of similarity existing between $\mu_Q(.)$ and $\pi_{A(x)}(.)$. In these models, both the query condition and the vague data stored in the fuzzy database are regarded as elastic constraints on the actual manifestation; then the degree of satisfaction of the query is computed as the degree of similarity between the two constraints.

In [3] a generalization of this second approach, named representation based approach, has been proposed in which one can use other measures than similarity to evaluate a vague query constraint in a fuzzy database.

For example, we suggest the use of a fuzzy inclusion measure to evaluate if a query constraint $\mu_Q(.)$ is satisfied by the attribute value $A(x)$ represented by $\pi_A(.)$. Specifically, the query satisfaction value is computed as the degree of a fuzzy inclusion of the vague attribute value $\pi_A(.)$ in the vague query constraint $\mu_Q(.)$ as follows:

$$Degree(A(x) \subseteq Q(x)) = \frac{|A(x) \cap Q(x)|}{|A(x)|} = \frac{\sum_{x \in D} min(\pi_A(x), \mu_Q(x))}{\sum_{x \in D} \pi_A(x)}$$

in which |A| is the cardinality of the fuzzy set A [18].

The use of a fuzzy inclusion measure instead of a similarity measure allows to define a flexible constraint based query language satisfying the following properties:

- When the actual attribute value certainly satisfies the query constraint Q, then the query satisfaction value is equal to 1: in facts, if $A(x) \subseteq Q(x) \Rightarrow \pi_A(x) \leq \mu_Q(x)\ \forall\ x \in D \Rightarrow A(x) \cap Q(x) = A(x)\ \forall x \in D \Rightarrow Degree(A(x) \subseteq Q(x)) = 1$. This is not the case by using a similarity measure since we can have query satisfaction values lower than 1 also in the case of certain satisfaction;

- If the actual attribute value certainly does not satisfy the query constraint Q, then $Degree(A(x) \subseteq Q(x))$ is equal to 0.
- Increasing query satisfaction values imply an increase of the possibility that the actual attribute value satisfies the query condition $Q(x)$; however in these cases there is uncertainty.

5.3   Representing Imperfect Spatial Information within fuzzy set and possibility theory

The problem of analyzing the nature of imperfection in spatial information has been pursued by many researchers [5,12,11,26,15,22,9,21,27]. Although there is an unanimous agreement in considering geographic information affected by different kinds of imperfection, up to date there is not a unique unambiguous vocabulary to identify the kinds of imperfection.

Worboys in [26] considers *imprecision* in spatial data as deriving by the granularity or resolution at which the phenomena have been observed and by the limitations of the computational representations. In his approach imprecision is distinguished from both *inaccuracy* leading to deviations from a real world value (errors and *inconsistencies*), and *vagueness* related to the inherent indeterminateness of geographic phenomena themselves.

In [22] the authors talk about *partial* spatial information as related to the resolution of the spatial data. In some applications such as the coordination of first-aid vehicles on a geographic region (taxi, metro etc.) each vehicle is represented as an *indeterminate* point on the region map since what is known about the vehicle position is its *scope* which is the area that it can reach in 2 minutes from his current position and the direction indicated by the driver by using terms such as *moving south station*.

Erwig and Schneider in [15] distinguish between spatial *indeterminacy* and spatial *vagueness* of geographic phenomena. They point out that there are two categories of spatial objects those with *indeterminate* boundaries, i.e. having sharp boundaries whose position and shape are unknown or cannot be measured precisely, and those with *not-well defined* boundaries or *vague* boundaries (examples are phenomena with continuously changing properties such as pollution, temperature, air pressure). In the first case we have uncertainty, in the second case we are in presence of *fuzziness*. Also Usery in [23] makes the distinction between features with fuzzy boundaries, that can be modelled by fuzzy subsets of the spatial domain, and features with determined boundaries but measurement uncertainty, that can be modelled as probabilities. Further, a number of other researchers have taken for synonymous terms *indeterminate* objects and *fuzzy* objects [5].

The objective of this section is to define some terms used for imperfect spatial information and to represent their semantics within possibility theory and fuzzy set theory. In facts, it is of basic importance to be able to distinguish whether we have indeterminacy or vagueness when modelling applications involving the representation of information. Depending on the identified type of imperfection we can then choose the more appropriate formal tool to deal with it.

We start from the assumption that we have an application and that we are going to represent the information on spatial data derived by an observation of a geographic phenomenon. Couclidis in [11] has pointed out that in representing imperfect spatial information in a database one should take into account three distinct and not unrelated things: the observed phenomenon itself, the characteristics of the observation, and the human intentionality or purposes. Imperfect information can derive by one or more of the above factors.

5.3.1   Inconsistency

It derives by the availability of multiple observations of the same phenomenon or when the observation is performed with inaccuracy. Inconsistent information may lead to contradictions and conflicts. It may occur when distinct thematic layers carry different information with respect to the same spatial unit. So, for instance, a forest inventory may indicate that a given land unit is covered by woods, whereas an agricultural map depicts the same unit as a vineyard. This is typically due to the fact that the forest and agriculture surveys are carried out by different agencies or departments, each one generating its own data set.

5.3.2   Fuzziness

It is related to the nature itself of the observed spatial phenomenon characterized by *gradual* physical, social, or cultural *properties*. Such phenomena can be represented by *fuzzy* spatial objects characterized by unsharp boundaries. Many natural spatial entities are fuzzy by definition: a typical example is represented by natural ecosystems, whose boundaries can often be set only on an arbitrary basis. Different forest types, indeed, are not sharply separated, but they gradually merge into each other.

The boundary of fuzzy spatial objects are characterized by a *gradual transition* from full satisfaction of the property to complete not satisfaction. In the literature several models have been proposed to represent fuzzy spatial objects [16,17,25,30]. In the approach in [30] a fuzzy spatial object A is modelled by a fuzzy set on a universe that is a spatial domain X, where the spatial domain X is defined as a bounded subset of $\Re^2$. The fuzzy set notation of A is the following:

$$A(x) = \int_{x \in X \subset \Re^2} \mu_A(x)/x \,|\, \mu_{A(x)} \in [0, 1]$$

in which $\mu_A(x)$ is the membership degree of element x to the fuzzy subset A.

The core is characterized by the spatial elements, that fully satisfy the property, i.e., $\{x \in X | \quad \mu_A(x) = 1\}$ the exterior by the elements that do not satisfy the property, i.e., $\{x \in X \quad | \mu_A(x) = 0\}$ and the boundary by the elements with partial degrees of satisfaction. More specifically, the boundary is constituted by α-cut level sets of spatial units with $0 < \alpha < 1$, i.e., $\{x \in X | \quad \mu_A(x) \geq \alpha\}$ such that α increases by moving from the exterior to the core; the spatial units belonging to each α-cut level set satisfy the property at least to the degree α.

Notice that in this case the real observed phenomenon is vague, and by means of fuzzy spatial objects we are trying to represent its natural vagueness. So, we have a conjunctive interpretation of the fuzzy set representing a fuzzy spatial object, since each element of the spatial domain is part of the real spatial entity to a given degree.

Based on the considerations that defining membership functions for fuzzy objects is often difficult or even impossible, and in most applications the observer (expert) himself is asked to express his judgement on the membership of a region to class (a soil class for example), we propose to allow the association with the distinct elements of the space domain of linguistic membership degrees such as *low, medium, high, full* to express distinct satisfaction levels of the gradual property defining the fuzzy object. The semantics of such linguistic values is specified by fuzzy sets defined on the domain of the gradual property characterizing the fuzziness of the spatial entity. In this way a fuzzy object is represented by a second order fuzzy subset of the spatial domain X:

$$A = \Sigma_{x \in X} \quad label \ /x$$

in which *label*$\in$ T(*Strength*)= { *none, low, medium, high, full* }. The elements x of the spatial domain X, from now on called grains, can be related with pixels in a raster representation, or with primitive geometric entities such as lines, polygons etc. in a vector representation. So, the expert associates the linguistic membership degrees with a finite number of grains.

Let us make an example. Let us assume that we have to represent on a land map the *wet lands* on the basis of several measures of the *water level* taken in some ground points over different periods of time. In some points these measures can greatly differ: in some periods of the year the soil can be dry while in other periods it can be watered. In a traditional crisp database we are faced with the decision of representing these points as belonging or not to the *wet land*. In the fuzzy representation we propose we can associate with each point a linguistic membership degree to the *wet land* that summarizes the information available at the points (see Fig. 5.2).

5.3.3   Indeterminacy

It depends on the characteristics of the observation and is derived by the inability or impossibility to precisely delimit a phenomenon that by its nature

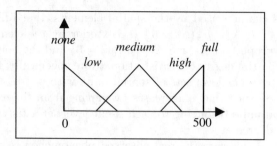

Fig. 5.2. Semantics of the linguistic membership values to a fuzzy object

can be crisp. Such characteristic is dependent on a lack of knowledge about the phenomenon so that we have uncertainty and vagueness in determining the real boundary or position of the spatial entity. Indeterminacy is strictly related to the granularity of the information, i.e., the level of details provided by the resolution of the spatial context below which spatial entities become undistinguishable. A common example of indeterminacy is represented by the use of land cover maps derived from the classification of satellite images. The spatial entities (e.g., the road network) distinguishable within such maps are obviously related to the spatial resolution of the images. To represent such phenomena indeterminate objects have been modelled in different ways: as spatial envelops [22] as objects with thick boundaries [8], or based on the egg-yolk model [10] just to cite some. In the context of possibility theory an indeterminate object A can be represented by a possibility distribution on a spatial domain:

$$A \subseteq \{x \mid \pi_A(x) > 0 \wedge x \in X\} with \pi_A : X \rightarrow [0,1]$$

The membership degree $\pi_A(x)$ of a grain x of the spatial domain X expresses the possibility that the spatial entity A covers the grain x: $\pi_A(x) = 1$ means that it is completely possible that the entity A covers the grain x; $\pi_A(x) = 0$ means that x is not a possible position for A; $0 < \pi_A(x) < 1$ means that there is a partial possibility for x to be covered by A. Notice that the observed actual entity can cover more grains of the space domain depending on the granularity of the observation; generally, its actual position will cover a subset of A. Indeterminacy can also affect the value of some spatial relationship (topological and metric relationships) between spatial entities (we name this kind of indeterminacy by relational indeterminacy). There are cases in which there is only a qualitative knowledge of the relationships between spatial entities expressed by linguistic terms such as region A is not completely inside region B, object A is mostly west of object B, and object A is very close to object B. A typical example is the determination of the position of a taxi-cab on a map knowing its indeterminate relationships with respect to well-known reference points on the map (ex. close to the north side of the railway station). In other cases, even if the precise knowledge is available, it is more convenient to manage a vague value for more efficient and effective query processing [9,19]. The linguistic values of an indeterminate distance relationship such as far, close, very close etc. can be represented within possibility theory by possibility distributions defined on $[0,1]$; the distance between two points on a reference spatial domain X has to be normalized with respect to the maximum possible distance on X in order to evaluate its degree of compatibility with the indeterminate distance values. The directional relationships can take values such as *south, south-west,* defined by possibility distributions on the domain $[0°, 360°]$ as in [19] (see Fig. 5.3). The topological relationship values such as *Disjoint, Overlap, Contains, Equal, Inside,* and their compound values such as *Almost Disjoint, Not Completely Inside, Almost Equal* etc. can be defined as possibility distributions on the unit range $[0,1]$. In the case of the symmetric topological relationships values *Disjoint, Overlap,* and *Equal,* and of their derived compound values the basic set $[0,1]$ can be defined as the domain of a similarity measure between fuzzy sets [18,7]. In the case of the asymmetric topological relationships values *Contains* and *Inside* and of their compound values the basic set $[0,1]$ can be defined as the domain of a fuzzy inclusion [18], see Fig. 5.3. In section 5.5, the rationale of these definitions of the indeterminate topological relationships will be discussed.

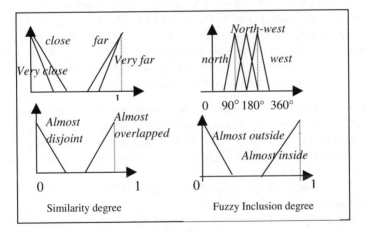

Fig. 5.3. Semantics of some linguistic values of indeterminate relationship between spatial entities

A further case of imperfection in spatial information involves both *fuzziness* (a vague phenomenon) and *indeterminacy* (due to the observation of the vague phenomenon). For example, the observation of a meteorological phenomenon such as the position of a *high* pressure zones. This kind of spatial entities can be represented as fuzzy sets of the second order defined on a universe of possibility distributions on the spatial domain X:

$$A = \sum_{\pi_{A_i} \in \wp(\pi)} label \, / \, \pi_{A_i} \, with \pi_{A_i} : X \to [0,1]$$

in which $label \in \mathrm{T}(Strength) = \{none, \, low, \, medium, \, high, \, full\}$ and $\wp(\pi)$ is the power set of the possibility distributions defined on the spatial domain X.

5.4   Representation of imperfect spatial information in a Fuzzy Object Oriented Database

There are two opposite data models for spatial information: the continuous *field based* model and the *exact object* based model [5]. While in the *field-based* model each attribute of a spatial phenomenon is assumed to vary continuously and smoothly over the space domain, in the *object-based* model the geographic entities have crisp boundaries and well-defined attributes. These two conceptual data models correspond with the raster and vectorial physical representation respectively.

These two models are inadequate for representing the real world spatial phenomena whose information is invariably affected by imperfection. For this reason some authors defined new conceptual models, others extended existing semantic models, mainly the Enhanced Entity-Relationship (EER) model, and the Object Oriented data base model [12,9,21,27]. In this section, by adopting a Fuzzy Object Oriented Database approach [2] we define new data types for representing the different kinds of imperfect spatial information identified in the previous section, specifically:

- the *Fuzzy Object Class* (FOC) whose instances are fuzzy objects representing spatial entities characterised by fuzziness;
- the *Indeterminate Object Class* (IOC) whose instances are indeterminate objects representing spatial entities whose position is partially or vaguely known;
- the *Fuzzy and Indeterminate Object Class* (FIOC) whose instances are fuzzy and indeterminate objects;
- the *Indeterminate Spatial Relationship Class* (ISRC) whose instances are indeterminate topological, directional and distance relationships between pairs of objects.

In this way we introduce a layer above the two conceptual models of spatial data which is functional to linguistically represent the imperfection related to spatial information (see fig. 5.4).

Further, a spatial entity can be naturally represented in an object oriented database context through a unique object having, besides its thematic and time attributes, the spatial reference attributes. In this way the one-to-one

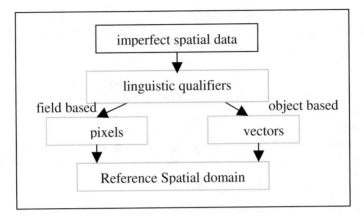

Fig. 5.4. Layers in the fuzzy object oriented database model for representing imperfect spatial data *depth*

correspondence with the application is maintained in the representation, and the spatial information becomes just one of the dimensions of the entity. In the following we will consider a spatial entity defined only by its spatial attributes, disregarding the thematic and time attributes.

5.4.1 Fuzzy Object Class (FOC)

The *Fuzzy Object Class* is defined in order to collect fuzzy spatial objects. Its intensional definition is the following:

Fuzzy Object Class definition:
{name:= *string*;
covers:={strength:=*strengthened property*;
spatial_reference:=subset of the spatial domain}
}

A FOC instance has a **name** whose domain is the set of strings and the multi-valued attribute **covers** that takes as value a set of pairs **(covers.strength, covers.spatial_reference)** with **covers.strength** defined as *strengthened property*, and **covers.spatial_reference** defined as a subset of a spatial domain. A spatial domain can be defined as either a grid of pixels in which each element is identified by its row and column indices or a subset of $\Re^2$ in which primitive geometric entities can be defined such as, points, polygons, etc. and identified by their co-ordinate values. The class *strengthened property* is defined so as to represent the semantics of the strength values with respect to the basic domain of the property:

strengthened property class definition:{
value:= label in *T(Strength)={none,low,medium,high,full}*;

basic domain:= numeric range;
semantics:= possibility distribution on the basic domain
}

An instance of strengthened property class has a property **name** that takes values in the string domain, a **value** that is a linguistic term in the term set of Strength, T(Strength)={*none, low, medium, high, full*}; a **basic domain** that is the basic domain of the strength property and the **semantics** that is a possibility distribution $\pi_{strength.value}$ on the basic domain of the strength property. For example, an instance o of the FOC is defined as follows:

o.name= *wet land*
o.covers.strength.name=*water level*
o.covers.strength.value=*full*
o.covers.strength.basic domain=*[0, 500]*
o.covers.strength.semantics= π_{full}
o.covers.spatial reference={(a,b)}
o.covers.strength.name=*water level*
o.covers.strength.value=*high*
o.covers.strength.basic domain=*[0, 500]*
o.covers.strength.semantics= π_{high}
o.covers.spatial reference={(a',b'), (c',d') }
o.covers.strength.name=*water level*
o.covers.strength.value=*low*
o.covers.strength.basic domain=*[0, 500]*
o.covers.strength.semantics= π_{low}
o.covers.spatial reference={(a",b"), (c",d"), (e",f") }

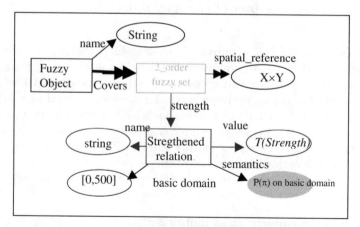

Fig. 5.5. Graphic sketch of the intensional definition of the FOC

In Fig. 5.5, a graphic representation of the intensional definition of the FOC is depicted. Rectangular shaped nodes identify structured classes, directed arrows identify attributes (multi-value attributes have a double arrowhead), oval shaped nodes identify simple attribute domains, i.e. domains that are built into the system such as string, integer, real etc. The oval shaded node labelled *"P(π) on basic domain"* identifies the set of all possibility distributions defined on the basic domain. In order to distinguish it from the simple domains it is shaded, so that we deal with it as a simple domain in this context, but we are made aware that its must be defined as a structured class.

In this example, **covers.strength.value** is used to summarize different water levels resulting from distinct observations (see fig. 5.2).

Formally the attribute covers defines a second order fuzzy subset:

$$o.covers := \{full/\{(a,b)\}, high/\{(a',b'),(c',d')\}, low/\{(a'',b''),(c'',d''),(e'',f'')\}\}$$

with (a,b), (a',b'), (c',d'), (a'',b''), (c'',d''), (e'',f'') pixels indices on a grid.

It is worthwhile to notice that a crisp spatial object can be defined as a particular instance of the FOC with the component covers.strength.value=*full* for all the instances of covers.spatial_reference.

5.4.2   Indeterminate Object Class (IOC)

The *Indeterminate Object Class* definition is the following:

Indeterminate Object Class definition:{
name:= *string*;
covers.spatial_reference:= possibility distribution on the spatial domain}

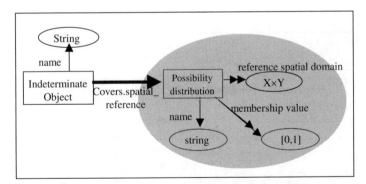

Fig. 5.6. Graphic sketch of the intensional definition of the IOC

In Fig. 5.6, a graphic representation of the intensional definition of the IOC is depicted. The property **covers.spatial_reference** takes as value the possibility distribution $\pi_{covers.spatial\_reference}$ defined on a spatial domain $X \times Y$. In this case, we do not know the position in the spatial domain of the represented spatial entity, but its possible position.

5.4.3   Fuzzy and Indeterminate Object Class (FIOC)

The intensional definition of the Fuzzy and Indeterminate Object Class is the following:

Fuzzy Indeterminate Object Class definition:
{name:= *string*;
covers:={strength:=*strengthened property*;
spatial_reference := possibility distribution on the spatial domain}

}

In Fig. 5.7, a graphic representation of the intensional definition of the FIOC is depicted. It can be noticed that it presents attributes of both the FOC and the IOC. Like in the case of the IOC, the attribute **covers.spatial _reference** is defined as a possibility distribution on a spatial domain. However, in this case we also have the attribute **covers.strength**. For example, we can define an instance o of the class FIOC whose name is *Wet land* and whose covers attribute takes as value the set of pairs: (*full*; *close*(a,b)); (*high*; *around*(a,b) in which *close*(a,b) and *around*(a,b) are possibility distributions. Formally, the attribute covers defines a second order fuzzy subset on possibility distributions with the meanings of the linguistic degree of membership *full* and *high* defined by the functions depicted in fig. 5.2.

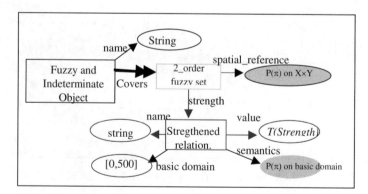

Fig. 5.7. Graphic sketch of the intensional definition of the FIOC

5.4.4   Indeterminate Spatial Relationship Class (ISRC)

Indeterminate spatial relationships between objects can be defined by means of the *Indeterminate Spatial Relationship Class* whose definition is the following:

Indeterminate spatial relationship class definition:{
type:= a value in {*distance, direction,topology*};
value:= linguistic term in the term set of type T(type);
basic domain:= $[0, 360]°$ | $[0, 1]$;
value semantics:= possibility distribution on the basic domain;
argument 1:= instance of FOC, IOC or FIOC;
argument 2:= instance of FOC, IOC or FIOC }

In Fig. 5.8, a graphic representation of the intensional definition of the ISRC is depicted. The relationship **value** is a linguistic term defined in the term set T(type); for example T(*distance*) can be defined as {*very close, close, far, very far*}; T(*direction*) can be specified as { *north, east, northeast, south, south-east, west, north-west, south west*} and T(*topology*) can be set as {*Overlap, Contains, Equal, Inside; Almost Disjoint, Not Completely Inside, Almost Equal, Almost Contains,* etc}. The choice of the term set of a relationship depends on the application. The **basic domain** definition depends on the type of the relationship: in the case of *distance* and *topology* relationships it is defined as $[0,1]$; in the case of *direction* it is equal to $[0°, 360°]$. The value of an indeterminate relationship is specified by a linguistic term (the **relationship value**) represented by a possibility distribution (the **relationship value semantics**). For example, in fig. 5.3 the relationship value semantics of some linguistic values are depicted.
argument 1 and argument 2 are instances of one of the classes FOC, IOC and FIOC.

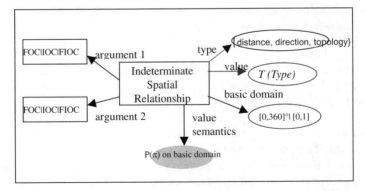

Fig. 5.8. Graphic sketch of the intensional definition of the ISRC

5.5 Vague query constraints evaluation on imperfect spatial information

In this section we trace the guidelines for the definition of a flexible spatial constraint-based query language for a fuzzy object oriented database containing fuzzy and indeterminate spatial objects.

As far as the non-spatial dimensions, the problem of evaluating vague constraints is the same as that faced in fuzzy databases and introduced in section 5.2. In this situation a vague constraint on a non-spatial attribute value can be expressed linguistically by a label and can be represented as a fuzzy subset on the attribute domain. Since also the thematic and time attribute values can be ill-defined, i.e. can be affected by indeterminateness, the flexible query constraints can be evaluated based on one of the methods summarized in section 5.2. We suggest the use of a fuzzy inclusion measure as described at the end of section 5.2. For example, *recent, ancient,* etc can be specified as vague constraints on the time attribute of a spatial entity and thus can be represented by fuzzy subsets of the time domain.

As far as the spatial dimension, there are three kinds of spatial relationship constraints that can be specified in a query: distance, directional and topological constraints. To allow flexible formulation of spatial constraints so as to produce discriminated answers we first have to define the vague spatial constraints semantics, then, the evaluation function of such vague spatial constraints between both crisp objects and fuzzy and indeterminate objects [4,20,19]. The most general situation involves fuzzy and indeterminate objects. An example of query specifying a vague distance constraint between fuzzy and indeterminate spatial objects can be expressed by the triple (*wet lands, close, coast*) that asks for the mapping of the *wet lands* instances that are *close* to a *coast* instance. Both the *coasts* and the *wet lands* spatial entities can be defined by means of the spatial data type FIOC and the distance relationship constraint is specified by the vague term *close* whose semantics is defined by a fuzzy set such as the one depicted in fig. 5.3.

Vague spatial constraints can be atomic or compound ones: a vague atomic spatial constraint is specified by a triple (a_1, C, a_2) in which:

- C is a linguistic term expressing the constraint; it is a value in the term set T(*type*) of an indeterminate spatial relationship. Its semantics is defined by a fuzzy subset μ_C such as one of those depicted in fig.5.3.
- In the most general formulation a_1 and a_2 are names of spatial entities such as *wet land* and *coast*, i.e., $a_1 = t_1$ and $a_2 = t_2$ with t_1 and t_2 name attribute of the spatial data types FOC, IOC and FIOC. By this formulation one specifies the evaluation of the constraint C between all pairs of instances o_1 and o_2 of the classes t_1 and t_2 respectively. A more specific constraint formulation is the one in which a_1 and/or a_2 are spatial objects identifiers, i.e., $a_1 = o_1$ and $a_2 = o_2$ with o_1 and o_2 instances of the spatial data types FOC, IOC and FIOC. For example, (*wet lands,*

inside, o_2) asks for the mapping of the *wet lands* that are *inside* the region identified by o_2. For example, o_2 can be drawn on the map by the user during the formulation of the query.

Compound spatial query constraints are defined as conjunctions of atomic ones [19].

The evaluation function of an atomic spatial query constraint (t_1, C, t_2) can be defined by first considering the case in which t_1 and t_2 are names of instances of the FOC, being crisp spatial entities particular instances of a FOC. During the query evaluation process, the value $f_C(o_1,o_2) \in [0,1]$ is computed and is interpreted as the degree of satisfaction of the C constraint by the (o_1,o_2) pair. In the following, we will define the evaluation function of vague topological query constraints. We remit the definition of distance and directional constraints evaluation functions between fuzzy and indeterminate objects to a future work.

We distinguish two kinds of topological constraints, the symmetric and the asymmetric constraints. A symmetric constraint has associated with it an evaluation function f_C that produces the same result by exchanging its arguments:

C is symmetric iff $f_C(o_1,o_2) = f_C(o_2,o_1)$

For example, the topological constraints *overlap, disjoint, almost disjoint etc.* are symmetric constraints.

C is asymmetric iff $f_C(o_1,o_2) \neq f_C(o_2,o_1)$

For example the constraints, *inside, outside, almost inside, contains etc.* are asymmetric constraints.

When C is symmetric its meaning μ_C is defined on the domain of a similarity measure S as specified in section 5.3.3 and then the evaluation function f_C is defined as:

$$f_C(o_1, o_2) = \mu_C(S(\underline{\alpha}, \underline{\beta})) \, with \, S : [0, 1] \times [0, 1] \to [0, 1] \, S(\underline{\alpha}, \underline{\beta}) = \frac{|\underline{\alpha} \cap \underline{\beta}|}{|\underline{\alpha} \cup \underline{\beta}|}$$

(5.1)

in which $\underline{\alpha}$ and $\underline{\beta}$ are fuzzy subsets of the spatial domain obtained by the second order fuzzy subsets $\alpha = o_1$.covers and $\beta = o_2$.covers defined in section 5.4.1, through a defuzzification of the linguistic membership values. $|\delta|$ denotes the cardinality of the fuzzy subset δ and μ_C is the function specifying the semantics of the C constraint [21]. If both o_1 and o_2 are crisp objects, then α and β are classic sets, and therefore $|\alpha|$ and $|\beta|$ can be defined as the areas of the spatial entities represented by the objects o_1 and o_2. By computing the similarity measure (5.1) we get values that increase as the two spatial entities represented by α and β are more overlapped. It gets the maximum value of 1 when α equals β. Based on these considerations, the meanings of the constraints *overlap,* and *almost overlap* will be defined as monotonic non decreasing functions in [0,1], while the meanings of the constraints *disjoint,* and *almost disjoint* will be defined as monotonic non increasing functions in [0,1] (see Fig. 5.3).

When the topological constraint C is asymmetric, its meaning μ_C is defined on the domain of a fuzzy inclusion I as specified in section 5.3.3. In the case of the constraints *inside, almost inside*, the evaluation function f_C is defined as:

$$f_C(o_1, o_2) = \mu_C(I(\underline{\alpha}, \underline{\beta})) \, with \, S : [0,1] \times [0,1] \to [0,1] \, I(\underline{\alpha}, \underline{\beta}) = \frac{|\underline{\alpha} \cap \underline{\beta}|}{|\underline{\alpha}|}$$

$$(5.2)$$

In this way, by applying (5.2) we get values that increase as more portions of α are included in β: the maximum value of 1 is obtained when α is completely inside β. Based on this consideration the meanings of the constraints *inside* and *almost inside* are defined as monotone non decreasing functions in [0,1], while the constraints *outside* and *almost outside* are defined by monotonic non increasing functions in [0,1] (see fig. 5.3).

In the case of the asymmetric constraints *contains, almost contains*, the evaluation function f_C is defined as:

$$f_C(o_1, o_2) = \mu_C(I(\underline{\beta}, \underline{\alpha})) \, with \, S : [0,1] \times [0,1] \to [0,1] \, I(\underline{\beta}, \underline{\alpha}) = \frac{|\underline{\beta} \cap \underline{\alpha}|}{|\underline{\beta}|}$$

$$(5.3)$$

In this way, by applying (5.3) we get values that increase as more and more portions of α contain β. The maximum value of 1 is obtained when α completely contains β. Based on this consideration the meanings of the constraints *contains* and *almost contains* are defined by monotone non decreasing functions in [0,1].

Now, let us consider the case of the vague atomic query constraints (t_1, C, t_2) where t_1 and t_2 are names of spatial entities defined as IOC. In this case $\pi_{o_1 covers.spatial\_reference}$ and $\pi_{o2 covers.spatial\_reference}$ (values of the covers.spatial_reference attribute of the objects o_1 and o_2) are possibility distributions on the spatial domain. In order to evaluate the C constraint, from $\pi_{o_1 covers.spatial\_reference}$ and $\pi_{o_2 covers.spatial\_reference}$ we first derive the two sets α_λ and β_λ as the sets of λ-level possible cover values of o_1 and o_2 with $\lambda \in [0,1]$:

$$\alpha_\lambda = \{x| \; \pi_{o_1 covers.spatial\_reference}(x) > \lambda\} \, and$$
$$\beta_\lambda = \{x| \; \pi_{o_2 covers.spatial\_reference}(x) > \lambda\}$$

with x grain of the spatial domain X. α_λ and β_λ represent the possible positions of the two spatial entities o_1 and o_2 with a possibility degree at least equal to λ. Then, we evaluate the C constraint's satisfaction by applying formula (5.1) , (5.2) or (5.3): in this case $f_{C,\lambda}(o_1,o_2)$ expresses the satisfaction degree of the C constraint by the two λ-level possible positions of the spatial entities o_1 and o_2. The same procedure is used to evaluate the constraint between entities defined as FIOC. In this case α_λ and β_λ are second order fuzzy subsets of the spatial domain and first we have to defuzzify the linguistic membership values before to compute the satisfaction degree of the constraint by applying formula (5.1), (5.2) or (5.3).

5.6   Conclusions

A linguistic modelling of vague and indeterminate spatial phenomena is well accepted by experts in various application fields. Based on a Fuzzy Object Oriented Database approach some spatial data types have been defined to represent vague and indeterminate spatial information. Then, the problem of the definition of a flexible spatial query language has been faced and a method to evaluate vague topological query constraints between fuzzy and indeterminate spatial objects has been proposed. This is a first step towards the definition of a new conceptual data model for managing imperfect spatial information that stands above the two opposed geographic data models the exact object based model and the field based model.

References

1. E. Bertino and G. Vantini. Advanced database systems and geographical information systems. In *II Workshop on AITSES*, pages 1–10, 1996.
2. G. Bordogna, D. Lucarella, and G. Pasi. A fuzzy object-oriented data model managing vague and uncertain information. *Int. J. of Intel. Syst.*, 14(7):623–651, 1999.
3. P. Bosc and O. Pivert. A formal framework for representation-based querying of databases containing ill-known data. In *2nd IIA'99 SOCO'99*, 1999.
4. P. Bosc and H. Prade. An introduction to the fuzzy set and possibility theory-based treatment of flexible queries and uncertain and imprecise databases. In A. Motro and P. Smets, editors, *Uncertainty Management in Information Systems: from Needs to Solutions*, pages 285–324. Kluwer Academic Pub., 1994.
5. P.A. Burrough and A.U. Frank. *Geographic Objects with Indeterminate Boundaries*. GISDATA series. Taylor & Francis, 1996.
6. R. De Caluwe. *Fuzzy and Uncertain Object-Oriented databases, Concepts and Models*. World Scientific, 1997.
7. T. Cheng and M. Molenaar. Diacronic analysis of fuzzy objects. *Geoinformatica*, 3(4):337–356, 1999.
8. E. Clementini and P. Di Felice. An algebraic model for spatial objects with indeterminate boundaries. In P.A. Burrough and A.U. Frank, editors, *Geographic Objects with Indeterminate Boundaries*, GISDATA series, pages 155–169. Taylor & Francis, 1996.
9. M. Codd, H. Foley, F. Petry, and K. Shaw. Uncertainty in distributed and interoperable spatial information systems. In G. Bordogna and G. Pasi, editors, *Recent Issues on Fuzzy Databases*, volume 53, pages 85–108. Physica Verlag, 2000.

10. A.G. Cohn and N.M. Gotts. The "egg-yolk" representation of regions with indeterminate boundaries. In P.A. Burrough and A.U. Frank, editors, *Geographic Objects with Indeterminate Boundaries*, GISDATA series, pages 171–187. Taylor & Francis, 1996.

11. H. Couclidis. Towards an operational typology of geographic entities with ill-defined boundaries. In P.A. Burrough and A.U. Frank, editors, *Geographic Objects with Indeterminate Boundaries*, GISDATA series, pages 45–56. Taylor & Francis, 1996.

12. V. Cross. Using the fuzzy object model for geographical information. 18th NAFIPS. pages 814–818, 1999.

13. D. Dubois and H. Prade. *Possibility Theory: an Approach to Computerized Processing of Uncertainty*. Plenum Press, New York, 1988.

14. D. Dubois and H. Prade. Tolerant fuzzy pattern matching: an introduction. In P. Bosc and J. Kacprzyk, editors, *Fuzziness in Database Management Systems*, pages 42–58. Physica Verlag, 1999.

15. M. Erwig and M. Schneider. Vague regions. In A.U. Frank and W. Kuhn e, editors, *Advances in Spatial Databases (SSD'97)*, volume 1262, pages 298–320. Springer Verlag, 1997.

16. Y. Leung. On the imprecision of boundaries. *Geographical Analysis*, 19:125–151, 1987.

17. D.M. Mark and F. Csillag. The nature of boundaries on "area-class" maps. *Cartographica*, 26:65–77, 1989.

18. S. Miyamoto. *Fuzzy sets in Information Retrieval and Cluster Analysis*. Kluwer Academic Pub, 1990.

19. D. Papadias, N. Karacapilidis, and D. Arkoumanis. Processing fuzzy spatial queries: a configuration similarity approach. *Int. J. of Geographical Information Science*, 13(2):93–118, 1999.

20. F.E. Petry. *Fuzzy Databases*. Kluwer Academic Pub., 1996.

21. W.V. Robinson. On fuzzy sets and the management of uncertainty in an intelligent geographic information system. In G. Bordogna and G. Pasi, editors, *Recent Issues on Fuzzy Databases*, volume 53, pages 85–108. Physica Verlag, 2000.

22. T. Topaloglou and J. Mylopulos. Representing partial spatial information in databases. *Geoinformatica*, 1998.

23. E.L. Usery. A conceptual framework and fuzzy set implementation for geographic features. In P.A. Burrough and A.U. Frank, editors, *Geographic Objects with Indeterminate Boundaries*, GISDATA series, pages 71–86. Taylor & Francis, 1996.

24. M.A. Vila, J.C. Cubero, J.M. Medina, and O. Pons. A conceptual approach for dealing with imprecision and uncertainty in object-based data models. *Int. J. of Intel. Syst.*, 11(10):791–806, 1996.

25. F. Wang and G. Brent Hall. Fuzzy representation of geographical boundaries in GIS. *Int. J. of GIS*, 10(5):573–590, 1996.

26. M.F. Worboys. Imprecision in finite resolution spatial data. *Geoinformatica*, 2:257–280, 1998.

27. A. Yaziki and K. Akkaya. Conceptual modeling of geographic information system applications. In G. Bordogna and G. Pasi, editors, *Recent Issues on Fuzzy Databases*, volume 53, pages 129–151. Physica Verlag, 2000.

28. L.A. Zadeh. Fuzzy sets as a basis for a theory of possibility. *Fuzzy Sets and Systems*, 1:3–28, 1978.

29. L.A. Zadeh. The concept of a linguistic variable and its application to approximate reasoning. parts I, II. *Information Science*, 8:199–249 and 301–357, 1997.

30. F.B. Zhan. Topological relations between fuzzy regions. *ACM SAC*, pages 192–196, 1997.

6 Using Classical Object-Oriented Features to Build a Fuzzy O-O Database System[*]

Fernando Berzal, Nicolás Marín, Olga Pons, and María Amparo Vila

Dept. of Computer Science and Artificial Intelligence,
Escuela Técnica Superior de Ingeniería Informática,
University of Granada,
18071-Granada-Andalucía-Spain
{fberzal,nicm,opc,vila}@decsai.ugr.es

6.1 Introduction

Real world data, as well as having more or less complex structures and relationships, can be affected by imperfection: we may be not very sure about the data certainty, or, even, data may be stated in a not very precise way. Whatever their formulation, data must be appropriately represented in databases and programs. Fuzzy Subset Theory[34] has proved to be an adequate tool to handle the different types of lack of precision and certainty that can appear in real world data.

The object-oriented data model is considered to be one of the most useful paradigms both in the database world [3] and in the programming field. Programmers and designers take advantage of its powerful modeling capabilities when they want to develop applications to solve problems characterized by a complex schema[4,11]. This model organizes the information as a set of objects with their own identity, grouped using the concept of class and type. This data conception is completed with important notions as the encapsulation of both the objects structure and their behaviour and the capability of building inheritance hierarchies with classes.

Due to this popularity among computer science professionals, and as it was previously made with other data models such as the relational one, many researchers try now to improve the classical object-oriented data model using fuzzy concepts. As a result, Fuzzy Object-Oriented Data Models (FOODM) have appeared in the literature[12].

Taking into account the ideas of [27], the addition of vagueness to improve the object-oriented database modeling can be considered at different levels:

- At the *attribute level*, dealing with the insertion of fuzzy data in the database by means of the use of fuzzy attribute domains.

[*] This work was partially supported by Spanish R&D project TIC99-0558 "Implantación de las Técnicas de la Computación Flexible en Aplicaciones de Gestión Empresarial. Aplicación a la Realización de un Módulo de Gestión de Oferta Inmobiliaria".

- At the *instance relationship level*, studying the fuzzification of the object membership to a given class.
- At the *inheritance relationship level*, softening the idea of the superclass-subclass relationships.
- At the *definition level*, considering the presence of fuzziness in the structure of the objects' type.
- And, finally, at the *behaviour level*, facing the fuzzification of the object conduct.

The first steps in the study of the addition of fuzziness to the object-oriented model are in closed relationship with advanced semantic data models, as the Fuzzy Entity/Relationship Model [23,35,28]. From the reflections opened by these works, many remarkable works have arisen, already in the object-oriented field.

In the early nineties J-P. Rossazza et al. [22] introduced a hierarchical model of fuzzy classes, explaining important notions (e.g. the tipicality concept) by means of fuzzy sets. A few months later, George et al. [9] began to use similarity relationships in order to model attribute values imperfection.

The main guidelines introduced by George et al. were extended by Yazici et al. [31,33,32] to develop a whole model during the past decade. By the middle nineties, Fuzzy Object-Oriented Database Models became an independent research field within the database research area. G. Bordogna and G. Passi [5,7] introduced an extended graphical notation to represent fuzzy object-oriented information based on the graph idea. At the same time, N. Van Gyseghem and R. de Caluwe [26,27] developed the UFO model, one the most complete proposals that can be found in the literature of this research area.

Other relevant works in this area can be consulted in [20,19,2,1,8]. Some of them define complex algebraic models while others are focused on the logic world, outside the scope of our work. Even the Entity/Relationship Model is currently being studied as a design tool for fuzzy object-oriented databases[13].

We have recently proposed a Fuzzy Object-Oriented Data Model[16,18,14] rooted on the idea of using classical object-oriented capabilities to implement fuzzy object-oriented extensions. We consider that the structures used to face the different types of data imperfection must be easy to understand and to manage, so that the effort required to implement a database system which incorporates them can be reduced. Ideally, a database designer without access to such a system, must be able to use the proposed structures in order to build effective solutions for her/his design problems.

In this chapter, we present the architecture of a fuzzy object-oriented database model built over an existing classical database system. Besides, we explain the different structures proposed in our model to handle data imperfection at the different levels it may appear, showing how to implement these structures using classical object-oriented features.

The chapter is organized as follows: Section 6.2 is devoted to the presentation of a general architecture for building a fuzzy object-oriented database system, using as its basis an existing classical object-oriented database system. Section 6.3 shows how to handle fuzzy extensions using classical object-oriented capabilities. Section 6.4 deals with some implementation issues that must be taken into account when developing a fuzzy object-oriented system like the one we propose here. Finally, some concluding remarks and future research trends are stated in section 6.5.

6.2   Building a fuzzy database system over a classical one

We can see in Fig. 6.1 the ANSI/SPARC standard architecture for any Database Management System in a simplified way. The lowest level of such architecture is formed by the *internal view* of the system, that is, the physical data definition and organization. At the other end of the diagram, at its highest level, we find the *external views* of the system, or, what is the same, the way the user (programmer of not) sees the database. The set of user or external views conforms the *external schema* of the database. Between these two levels, there is a *conceptual level* that represents the abstract view of the database according to the data model that characterizes the system.

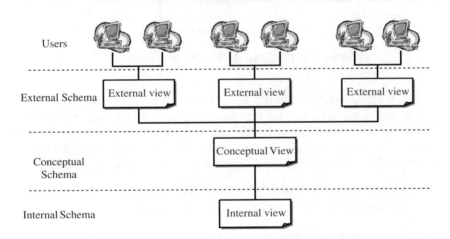

Fig. 6.1. ANSI/SPARC architecture

In our context, this architecture corresponds to a classical object-oriented database system and we want to modify it in order to obtain a Fuzzy Object Oriented Database System. Such modification implies the following changes:

- The external views are organized so that the user manages fuzziness transparently from the different structures used to implement it at lower levels. This will be the *fuzzy face* of the system.
- The conceptual schema will be organized into two sublevels: one of them will use fuzzy concepts to describe the conceptual representation of the database, while a lower-abstraction level will store the classical object-orientedv counterpart used to implement the structures and objects present in the higher one.
- The internal schema will be the appropriate for the classical system used to implement the FOODB.

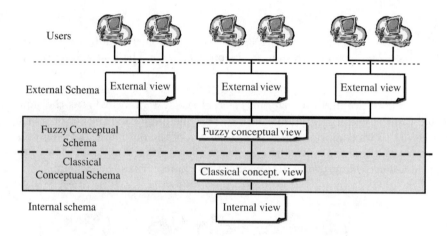

Fig. 6.2. Extended ANSI/SPARC architecture

Fig. 6.2 graphically shows this idea. If we want to develop a fuzzy database system using an existing classical one, the modifications presented in the figure lead us to an architecture organized in three levels. Fig. 6.3 explains these three levels in detail. As can be observed, the Fuzzy Object-Oriented Database Management System is founded on three fundamental pillars:

- The underlying classical database system, which will support most of the management functions and will store the information provided by the user.
- The conceptual fuzziness handler, built as a layer over the previous one, which will increase the capabilities of the underlying system, implementing the structures which are necessary to offer new mechanisms for data imperfection handling.
- An interface which allows the user to interact with the conceptual fuzziness handler layer, hiding the underlying layers and allowing users to develop and to manage their fuzzy object-oriented schemata and databases.

The users will be able to insert imperfect data when necessary, and will be able to design their applications using the new fuzzy capabilities of the fuzzy system.

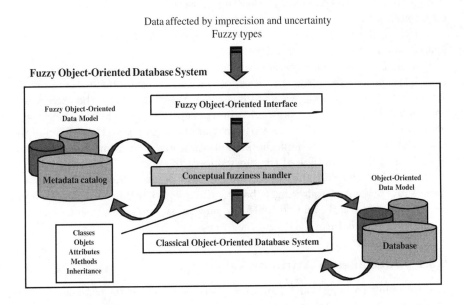

Fig. 6.3. System architecture

Data and metadata persistence is organized into two storage areas:

- A metadata catalog, which will store the fuzzy schema defined by the user, as well as all the information necessary to manage it from the database system point of view.
- A classical database, where the system actually stores the information provided by the user.

Taking this architecture into account, the next step is to develop the way to face fuzzy extensions for data imperfection management using classical object-oriented features. The following section is devoted to this topic.

6.3 Managing fuzzy capabilities using classical object oriented features

This section constitutes the fundamental part of our proposal. Here we analyze the presence of imprecision and uncertainty at all the levels they can appear within an object-oriented database model. First, we will focus on the attribute level, studying the different natures that imprecision can have when it affects attribute values and the way to face the lack of certainty about their veracity.

With respect to the instance relationship level, we show how to represent fuzzy extents for the classes, interleaving a discussion of the interpretation of such extents from a semantic point of view.

At the inheritance-relationship level, we focus our study towards generalization and specialization processes.

We make some considerations about uncertainty at the definition level, that is, when it affects the definition of the objects and classes themselves. Also in the definition level, we introduce a new concept of type, the *fuzzy type*, as a generalization of the classical type concept. We present this new type concept making an analysis both from the structural and the behavioural points of view. We also adapt the instantiation and inheritance mechanisms in order to take advantage of the new concept of type.

Remember that all these fuzzy extensions of the object-oriented data model are presented following a basic principle: to tackle uncertainty and imprecision separately and to build the new fuzzy extensions using classical object-oriented features.

6.3.1   Vagueness in Attribute Values

We are going to begin our vagueness study considering its presence in the attribute values that characterize objects. In our opinion, the capability of handling imperfect values in attributes must be built taking into account the different semantics the domains of such values can have.

Previous models, like the ones of George et al.[9] and Yazici et al.[31,33,32], introduce a clear distinction between the conjunctive and disjunctive semantics that attribute values can have, whether they are expressed using fuzzy sets or not. At the same time, Bordogna et al.[7] allow the use of labels to tag values which are presented using possibility distributions and linguistic variables.

We have taken positive ideas from these works to elaborate a whole model for the imperfect information management at the attribute level: we use the linguistic label expressiveness to handle imprecise values, considering the different semantics associated to the labels according to the characteristics of the domain over which they are defined. At the same time, we make a clear distinction between the representation of imprecision and that of uncertainty.

Taking into account the ideas of Ruspini [23], three different types of imperfect information may appear in an attribute value:

- The value is imprecise.
- The value is affected by uncertainty.
- The value is imprecise and affected by uncertainty at the same time.

Imprecise Attribute Domains

The reasons why an attribute value can be ill-defined may be assorted, from an actual ill-knowledge of the datum, till an essential imprecision affecting the attribute domain. For instance, let us consider the following examples:

1. Mary is in her *final* academic years.
2. Mary's mark is *good*.
3. Mary's prospects are *good*.

There exists a perfectly defined attribute domain in the two first sentences (i.e., a given number of academic years and a certain mark range). However, the last sentence is different because we cannot find an underlying domain to express the label *good* in a natural way. Therefore, in order to deal with imprecise information, we have to distinguish between these two cases, namely, with and without underlying semantic representation.

Domains without underlying semantics

This case corresponds to the third sentence "Mary's prospects are *good*". In these situations, the attribute domain is a set of linguistic labels, but we cannot define a semantic interpretation of such labels by means of fuzzy subsets over an underlying domain. To overcome this difficulty and to adequately manage the imprecision represented by the set of labels, we use a similarity relationship. Multiple values can be added to the domain using a new label and modifying the resemblance relation in a suitable manner [30].

Let D be the attribute domain (set of labels), S a similarity relationship defined over such domain, and L a fuzzy subset of D that represents the multiple value we want to include in the domain. To extend the domain, we have to take the following steps:

- Adding to D a new label l to represent the new value.
- Modifying S, so that $\forall x \in D, \mu_S(l, x) = max_{y \in D}(\mu_S(y, x) \otimes \mu_L(y))$.

Domains with underlying semantics

Let us now see how to face the representation of Mary's mark and academic year. In these two examples, there exists an underlying domain that allows us to define a fuzzy set to express the semantics for the linguistic labels. However, we must take care about the meaning given to their semantics since it may happen that the real attribute value can take only one of the values included in the distribution (e.g. the mark of Mary) or be a combination of such values (e.g. her academic year):

- If we have a possibility distribution with disjunctive interpretation (e.g. "Mary's mark is *good*"), the situation is very similar to the case with no underlying semantics from the domain structure point of view. The attribute has an atomic domain conformed by the union of a basic domain and a suitable set of labels. We also use a similarity relationship in the domain, but now it is defined as an extension of the classical equality, since we have a basic domain whose values must be taken into account [30].

Consider an attribute that can take value in a basic domain B and in a set of labels L whose semantics is expressed by means of fuzzy subsets of B ($\forall l \in L, \exists \mu_l : B \rightarrow [0,1]$). To manage the real domain of the attribute, we follow the next steps:

- Defining the actual domain as $D = B \cup L$.
- Building a resemblance relation S in D as follows:

$$\forall x, y \in D, s(x,y) = \begin{cases} 1 & (x = y) \wedge (x, y \in B) \\ 0 & (x \neq y) \wedge (x, y \in B) \\ \mu_l(z) & ((x = l \in L) \wedge (y = z \in B)) \\ & \vee ((y = l) \wedge (x = z \in B)) \\ max_{z \in B}(\mu_x(z) \otimes \mu_y(z)) & otherwise \end{cases}$$

(6.1)

- Otherwise, we deal with cases like the sentence "Mary is in her final academic years". That is, we have an attribute whose value is a fuzzy subset of a specific basic domain with an implicit conjunctive meaning. In these situations, whether we use a label to express the multiple value or not, we have to deal with a suitable representation of the fuzzy set.

For example, if the fuzzy set has a finite support-set, we can consider that the membership function is a finite set of pairs. That is:

$$\forall A \in \tilde{\mathcal{P}}(D) \mid Support(A) \text{ is finite}, A \subseteq \mathcal{P}(B \times [0,1]) \qquad (6.2)$$

If the support-set is not finite, we have to consider alternative representations for the membership function. In this case, we can use classical simplifications as triangular and trapezoidal functions.

Uncertainty in Attributes Values

Data, besides being imprecise, may be uncertain (in an explicit sense). Think about those situations where data is perfectly defined, but we are not sure whether they are true or not (e.g. "it is very possible that Mary lives in Granada"). A degree of truth must be considered for these situations.

Different scales can be used to express the lack of certainty about attribute values. For example, we can use probability measures (w.r.t. possibility) defined within the [0,1]-interval, or use linguistic probabilities (w.r.t. possibilities) whose labels may, in turn, have a semantic representation as a disjunctive fuzzy set at interval [0,1].

Considering that some kind of uncertainty is associated to an attribute value implies to assume that the actual attribute domain is an aggregation of the attribute domain and the scale where this uncertainty is measured. Independently of that fact, in case either the actual domain or the scale (or both) are fuzzy, the corresponding classes must be previously defined using the strategies explained above.

If we have imprecision and uncertainty at the same time (e.g. "It's possible that Mary's propects are good"), care must be taken to compute the similarity. This problem is complex and in some cases is not appropriately solved[6,10].

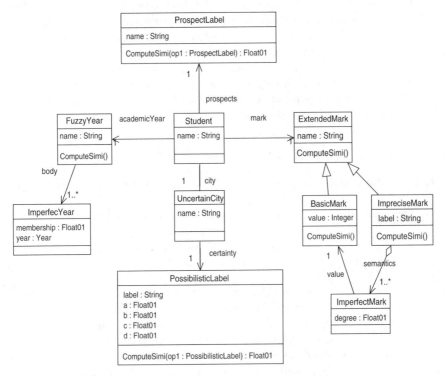

Fig. 6.4. Mary's example using the Unified Modeling Language[21]

Representation in classical object-oriented systems

Fig. 6.4 shows a complete example where imperfection is managed when appearing at the attribute level. We have taken into account the following considerations:

- In order to describe domains without underlying semantic, we define a class with an attribute to express the label, and a method capable of computing similarities between labels. In the figure, class *ProspectLabel* is implemented to represent the prospects of a given student.

- In order to describe domains with disjunctive meaning, we define an abstract class, which represents the actual domain, with two subclasses: one to represent the basic domain and other to represent the set of possible imprecise terms. The hierarchy of *ExtendedMark* has been added in the schema of Fig. 6.4 in order to express the mark of a given student.
- In order to describe domains with conjunctive meaning, we use a class to express the semantic of the fuzzy set. In fig. 6.4, we define a class *ImperfectYear* to express the pairs of the membership function. A class *FuzzyYear* with an attribute to express the label and other to express the semantic (a set of pairs) is used to describe the definitive domain.
- Finally, in order to express uncertainty, we construct the domains as aggregations of the corresponding domain and the scale used to express the lack of certainty. In the example, we consider that the attribute *city* of a given student can take values of class *UncertainCity*, which has been built following the previous guidelines.

6.3.2   Relationships among objects

Semantic data models usually offer two ways for connecting objects:

- The concept of attribute, which involves a functional approach and is used to relate a class with the domains of its attributes.
- The aggregation capability, which is used to model those explicit relationships not included in the previous case.

Suppose that we have two classes, and we want to establish a relationship between them. One alternative is to add to one of the classes (or to both of them) an attribute which points to the other class. The other alternative is to create a new class to express the relationship, with two attributes, each of them pointing to one of the classes that we want to relate. The application designer must decide in each moment how to manage the relationships in the schema, choosing between these alternatives.

The previous subsections have been devoted to the study of data vagueness representation when we use a functional approach. However, if the relationship is represented by means of an explicit class some appreciations must be taken into account:

- If we have partial knowledge about whether two objects are related, we can include a new attribute in the class standing for the truth level of the corresponding association, using the different truth scales we have previously commented.
- We can also consider that the connection admits importance degrees[7]. This case arises when not all the relationship instances have the same strength. In these situations, we can use numerical or linguistic levels to express this strength. Regarding linguistic scales, it should be noted

that labels should have a different meaning from that used to express uncertainty. For example, we could consider the set {"high", "medium", "low"} with each label represented as a disjunctive fuzzy set in [0,1].

6.3.3   Instance Relationship Level: fuzzy extents

The presence of imperfection in the attribute values that characterize any object, can make gradual the membership degree of an object to a given class. That is, the object belongs to the class with a membership degree within interval [0,1]. The class does not have a crisp extent of objects, but a fuzzy one.

There exist in the literature several proposals to compute the membership degree of an object to a given class, most of them founded on concepts such as inclusion and typicality of attribute values with respect to the expected ones or prototypical ones of the class[9,31,22].

The representation of this feature in the object oriented model is very simple: we only have to add an extra attribute in the class with fuzzy extent. Normally, this attribute should be [0,1]-valued, although considering the use of a set of labels might result interesting too. Constructor methods can be defined implementing the policy used to compute the membership degree.

The gradual membership of objects to a given class may not be only caused by the presence of imperfection in attribute values. It can also be due to a semantic necessity. Fig. 6.5 depicts the UML diagram for a class of Experts with fuzzy extent: a person is considered to be an expert with a certain degree.

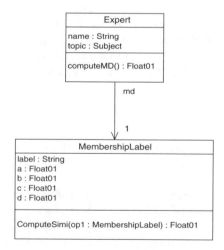

Fig. 6.5. Expert's example

6.3.4   Inheritance Relationship Level

Inheritance hierarchies can be built following two approaches: a top-down one, called specialization, and a bottom-up one, called generalization.

Fuzzy Subclasses

Specialization processes create subclasses from existing classes by constraining the description of a property (e.g. an attribute domain) or by specifying a set of additional properties.

Both kinds of mechanisms could lead to imprecise structures by considering flexible ways for characterizing the corresponding subclasses. This flexibility results in the fuzzyfication of the membership degree of each possible object to the subclass, which can be represented taking into account the guidelines stated in section 6.3.3.

When the subclass is described by constraining the description of a property, we actually have an intensive class in the sense of [29], since it is defined by means of a fuzzy predicate. Therefore, it will not be necessary to represent it as a separate structure and a suitable procedural management will suffice in most cases.

The creation of an explicit class to represent the subclass is only necessary in those situations where new properties are added and when the subclass takes part in new relationships.

For example, let us consider a subclass *GoodStudent* from the class *Student*. In order to accomplish the creation of such a class we only need to compute some heuristic rule over the student's marks and then to determine the student goodness degree. An extra attribute *goodnessLevel* in the Student class will suffice.

On the other hand, let us consider the creation of a subclass *Employee* from an existing class *Person*. Extra properties would characterize the new class, but at the same time the kind of job may result in a fuzzyfication of the extent of the class. This last case is similar to the *Expert* example. Fig. 6.6 shows both cases.

Fuzzy Categories

We can construct a superclass by means of the union of different existing classes, following a generalization process. In such a situation, any subclass object must belong to the created superclass.

Additionally, some semantic data models use the *category* concept to express the fact that a class is defined as a subset of a superclass. The definition process of a new category can be softened by considering that categories are fuzzy subsets of the involved class union. Let us see an example.

Suppose that we have two classes, namely, *Student* and *Teacher*. We can create the fuzzy category *Researcher* from these two classes: any student or teacher could be considered a researcher with a certain degree.

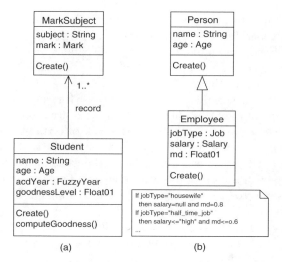

Fig. 6.6. Subclass creation

The representation of a fuzzy category in an object-oriented context involves the creation of a class with fuzzy extent. That is, an extra attribute is included together with some integrity constraints concerning the inclusion of the fuzzy category in the union of the initial classes. In the latter example, we have to respect the following constraint with respect the degrees:

$$\mu_{Researcher}(o) \leq \mu_{Student}(o) \vee \mu_{Teacher}(o).$$

6.3.5   Definition Level

We have studied the presence of vagueness from the data point of view. However, we have not considered the problem that arises when vagueness directly affects to the definition of both objects and classes. In this subsection we will discuss the sense and the representation of uncertainty in the schema, and we will present a new concept of type for managing different precision levels in the definition of the class type.

Uncertain Definitions

We can distinguish between two different situations with respect to uncertainty affecting the definition of a given object:

- When we are not sure about the value of some attributes: this situation corresponds to the attribute level, and does not actually involve any changes in the definition of the object.
- When we are not sure about whether an object has or has not a certain attribute. This fact concerns the object definition and has to be managed in a suitable manner.

The direct way to manage the uncertainty about the application of an attribute (or, generally, a property) to a given object is to consider an extra attribute that informs us about the applicability degree associated to each attribute. The scale of applicability can be, as usual, of several types (possibilistic, probabilistic, etc.).

Care must be taken in these situations because it results quite hard to distinguish the semantics of this uncertainty: in some situations it could be very difficult to make a distinction between the lack of certainty about whether an attribute characterizes an object or not and the lack of certainty about the reliability of the source that provides us the value for the attribute.

Besides, many of the situations where the use of uncertainty in a property applicability seems to be necessary may be due to an inappropriate definition of the class.

In the previous lines, we have considered the presence of uncertainty in object definition. Let us devote some lines to discuss the presence of uncertainty in the class definition.

Most situations where uncertainty affects class definitions appear during the initial steps of the modeling process, when we are not sure about the class structure yet.

This kind of lack of certainty involves important drawbacks:

- If we create an object of a class defined with uncertainty, this uncertainty must be propagated to the object definition and to the object membership degree itself. That fact makes data manipulation much more complex.
- With time, this uncertainty might change and this change might affect not only the schema but the objects in the database.

Thus, the presence of uncertainty in the class definitions must be avoided. As a matter of fact, one of the most important designer's aims is to eliminate uncertainty in the schema, trying to find the hierarchy of classes that best represents the problem to be solved.

Fuzzy Types

In the previous paragraphs we have discussed the unsuitability of uncertain structures, and we have concluded that this kind of vagueness at the definition level is due to an *ill* knowledge of the schema and the objects. Conversely, a good knowledge of the problem could lead us to manage different levels of precision in our structures. The following paragraphs are devoted to the study of this idea.

The structure associated to a given class can be viewed as a set of attributes or properties with a series of associated ranges. This concept of structure, that we call *crisp structure*, fulfills a large portion of the needs related to types when the hierarchical structure associated to a given application is being built.

However, there exist problems where this concept of structure is not suitable and a softening process is desirable. Examples of these problems are the representation of concepts with different levels of precision, semistructured and unstructured data management, representation of presence-absence attributes, and incomplete information handling. This kind of problems requires the use of more expressive and powerful techniques to define the structure of a certain class of objects.

In [16] we presented a new concept of type that comes to solve some of these problems. Let us now look at a brief summary of this concept and its most important characteristics. In order to discover more about the problems where fuzzy types can be useful, and obtain a wider presentation of the concept, see[16,18].

Structure and behaviour of a fuzzy type

Our new concept of type is founded on the idea of fuzzy data structure. Let us see its definition.

Definition 1 (Fuzzy structure) *A fuzzy structure is a fuzzy set defined over the set of all the attributes possible in our model.*

Taking this definition into account, we can define a fuzzy type as follows.

Definition 2 (Fuzzy type) *A fuzzy type is a type whose structural part S is a fuzzy structure.*

Let T be a type associated to a given class C. The membership function that characterizes the structural component of the type has the following form:

$$\mu_S : Attributes -> [0, 1] \tag{6.3}$$

where *Attributes* is the set of all attributes that can be used in our model.

The support set of any user-defined structure is always finite (it is meaningless to consider structures with an infinite support set) so the latter function can be expressed by means of the following simplified notation:

$$S = \mu_S(a_1)/a_1 + \mu_S(a_2)/a_2, ..., + \mu_S(a_n)/a_n \tag{6.4}$$

The set of attributes that can be used to characterize the type at any moment is the support set of the fuzzy set associated to the type. The kernel-set contains the basic attributes of the type. Finally, each one of the α-cuts defines a precision degree with which the type can be considered.

Let us use an example to clarify the previous definitions. Suppose that we want to define the concept of Image, and we use the following three levels of precision:

Object image
- *Minimum features*: theme, file, format and version.
- *First level of precision*: horizontal resolution, vertical resolution, and palette.
- *Second level of precision*: histogram, frontiers, bands, and convolution.

The structure of the type Image can be expressed using the following fuzzy set: S=1/theme + 1/file + 1/format + 1/version + 0.9/hor_res + 0.9/ver_res + 0.9/palette + 0.8/bands + 0.8/histogram + 0.8/convolution + 0.8/frontiers.

This set has three relevant α-cuts:

- S_1={theme, file, format, version}
- $S_{0.9}=S_1\cup$\{hor_res, ver_res, palette\}
- $S_{0.8}=S_{0.9}\cup$\{histogram, convolution, frontiers\}

When the user of our type wants to create an instance of the class Image, he will be able to make it incorporate the attributes of either S_1, $S_{0.9}$, or $S_{0.8}$, according to the precision degree required by the object to be represented.

The structural fuzziness of the type also affects the methods that define the behavioural component of the type of a class, because they include in their code references to attributes of the object over which they are applied.

So far, in the object-oriented model, every instance of a class could reference any of the attributes of the class (instance variables). However, with our new kind of types, an instance of a given class may not incorporate certain attributes depending on the α-cut of the class structure it has been created with.

Each one of the methods defined in a class must have an associated precision level (as is the case with the attributes or instance variables) that indicates the minimum precision that an instance must have to incorporate a method in its behaviour. This level of precision depends on the attributes and methods referenced in the method implementation.

Definition 3 (Precision level of the method) *We define the precision level in the code of a method m (NC_m) as the minimum of the membership degrees which have the attributes that directly appear in the code of the method. It and can be calculated as follows:*

$$NC_m = \begin{cases} 1 & \text{if } Attr(m)=\varnothing \\ min_{a\in Attr(m)}\, \mu_S(a) & \text{otherwise} \end{cases} \tag{6.5}$$

where $Attr(m)$ is the set of attributes referenced in the code of the method m. Taking the previous definition into account, the precision level (N_m) of a method m, that is, the degree it must have, can be calculated by means of the following formula:

$$N_m = \begin{cases} NC_m & \text{if } Ref(m)=\varnothing \\ min(min_{x\in Ref(m)}\{NC_x\}, NC_m) & \text{otherwise} \end{cases} \tag{6.6}$$

where $Ref(m)$ is the set of methods (directly or indirectly) referenced in the code of the method m (excluding recursive references).

In other words, in order to apply the method m, an object of the class C must have been created using an α-cut of the structure of the class C with α less or equal to N_m. Otherwise, the object would lack some necessary attributes to apply the given method.

For example, regarding to the type Image, it is possible to find methods with precision level 1 (as get_file(...), change_theme(...), etc.), with precision level 0.9 (as change_palette(...), etc.) or even with level 0.8, which generate image convolutions or histograms. That is, images created with α equal to 1 cannot apply a method to display their convolution.

Definition 4 (Behaviour of a Fuzzy Type) *The behavioral component B of a Fuzzy Type is a fuzzy set defined over the set of all methods that can be defined in our model whose membership functions is the following:*

$$\mu_B : Methods \rightarrow [0,1]$$

$$\mu_B(m) = \begin{cases} 0 & \text{if } m \text{ has not been defined for the class} \\ N_m & \text{if } m \text{ has been defined for the class} \end{cases} \tag{6.7}$$

where Methods is the set of all the methods that can be defined in our model.

Instantiation of fuzzy types

The change proposed in the concept of type involves modifications to the idea of instantiation.

In order to create a new object of a given class, we must be able to choose the α-cut of properties of the type that will be used to represent it. With that aim in mind, our model includes a generic method *new(α)* (with $\alpha \in (0,1]$). The receptor of this method can be any class C, while the parameter α is the level α of the structure that is needed to represent the new object.

The effect of sending the message *new(α)* to a class C with structural component S and behavioral component B, consists of creating an object incorporating the set S_α of attributes. The set B_α of methods defines the behaviour of this object.

For example, with respect to the class Image, the message Image.new(1) would create an Image object represented by the attributes {theme, file, format, version} and incorporating in its behavior the methods necessary to handle these attributes. Any α less than 1 and strictly more than 0.9 would have the same effect. Lower α values would create more precise objects, that is, objects characterized by a bigger set of attributes.

Fuzzy type inheritance

In order to take advantage of fuzzy types in our model, we propose mechanisms to deal with inheritance relationships in this kind of types.

The inheritance mechanism H must enable part of the class structure and behavior to be inherited by subclasses. As we have done with the instantiation mechanism, we add a threshold to indicate which subset of properties we want to be inherited. Two different forms of inheritance can be considered:

- Incorporating inherited attributes and methods to the kernel set of the structural and behavioural components of the subclass, respectively. In this way, the vagueness of the inherited properties will be eliminated. This type of inheritance will be called *inheritance without vagueness propagation* H_{crisp}.

  In the context of images, we might create a subclass *Plan* from *Image* with a crisp structure, incorporating all the properties of the first two levels of Image, but now with precision level 1.

- Keeping the vagueness, by inheriting both properties and methods affected by the corresponding membership degree. This type of inheritance will be called *inheritance with vagueness propagation* H_{fuzzy}.

  For example, we might create a subclass *Bio-Image* with level 0.8 to represent biological images. It might be also organized into three levels of precision, although some specific properties of this kind of images would be added to each inherited level.

Let us suppose that we are creating a new class C. We would like it to inherit from another class C' without vagueness propagation with a threshold α. S_{Sub} and B_{Sub} stand for the structural and behavioural components defined for the new class. S_{Sup} and B_{Sup} stand for the respective components of the superclass. The resulting type for the class C will have the following components:

$$S_C = S_{Sup_\alpha} \cup S_{Sub}, \tag{6.8}$$

$$B_C = B_{Sup_\alpha} \cup B_{Sub}. \tag{6.9}$$

However, in order to inherit with vagueness propagation, the new components would be:

$$S_C = S^*_{Sup_\alpha} \cup S_{Sub}, \tag{6.10}$$

$$B_C = B^*_{Sup_\alpha} \cup B_{Sub}. \tag{6.11}$$

In these last formulae, for any fuzzy set Y, Y^*_α stands for the fuzzy cut at level α of Y, defined by the following membership function:

$$\mu_{Y^*_\alpha}(m) = \begin{cases} 0 & \text{if } \mu_Y(x) < \alpha \\ \mu_Y(x) & \text{otherwise.} \end{cases} \tag{6.12}$$

In both kinds of inheritance, B_{Sup_α}, $B^*_{Sup_\alpha}$ and B_{Sub} are recalculated once the definitive structural component S_C is known.

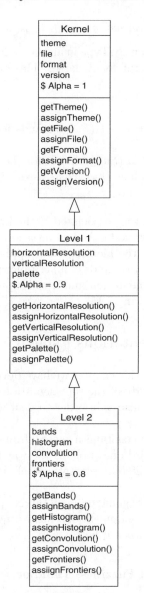

Fig. 6.7. Fuzzy Type Image

Representation in a classical object-oriented system

In order to represent a Fuzzy Type in an classical object-oriented database, we can use a 1-ramified hierarchy of classes. This kind of hierarchy is defined as a series of classes C_1, ..., C_{i-1}, C_i, C_{i+1}, ..., C_n verifying the following properties:

- $Sub_{Ci} = \{C_{i+1}\}$, $i = 1..n - 1$ (Sub_{Ci} stands for the set of subclasses of C_i).
- $Sup_{Ci} = \{C_{i-1}\}$, $i = 2..n$ (Sup_{Ci} stands for the set of superclasses of C_i).
- A finite sequence of values $\{\alpha_i\}$ exists associated to the hierarchy, such that $\alpha_1 = 1$, $\alpha_n > 0$ and $\alpha_i > \alpha_{i+1}$.

Fig. 6.7 shows the hierarchy associated to the Image type. Using this kind of hierarchies we can easily model the instantiation mechanism (we only have to chose the right class of the hierarchy and apply the usual *new* method). The implementation of the two kinds of inheritance is a bit more complicated. The details of such implementation and a wider presentation of the concept of fuzzy type can be found in[17,15].

6.4   Some implementations issues

In the previous sections of this chapter we have developed the general goal of our research effort: the study of the representation of fuzzy extensions using classical object-oriented features in order to deal with several kinds of data imperfection.

While the second section contains a general framework to develop a system following this idea, the third one explains the way to represent imperfection when it appears at the different semantic levels in a fuzzy object-oriented data model.

Having done that, we are going to finish this chapter by making some more appreciations about the development of a fuzzy object-oriented database system using the guidelines exposed here.

6.4.1   The Conceptual Fuzziness Handler

The central role in the architecture presented in Fig. 6.3 is played by the Conceptual Fuzziness Handler. This architectural layer is the most complex one because it establishes the correspondence between fuzzy data and types and the set of classical classes and objects that support them. That is, this handler must implement the new fuzzy capabilities using the tools provided by the underlying classical system.

Fig. 6.8 depicts this component structure. As can be observed in the figure, the Conceptual Fuzziness Handler splits its task between two subordinate handlers, namely, the Schema Handler and the Objects Handler:

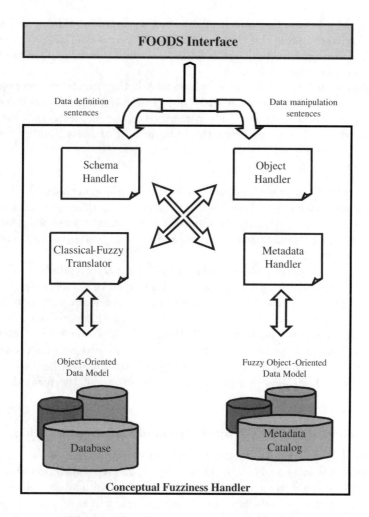

Fig. 6.8. The Conceptual Fuzziness Handler

- The Schema Handler solves the commands that come from the interface in relation with structure definitions in the database (mostly Data Definition Language related).
- The Object Handler allows the creation, storage, management, and deletion of objects over the structures created by the previous handler (Data Manipulation Language).

While the Schema Handler gives static support to the database, the Object Handler provides dynamic support, by managing the objects' lifecycle. Below these handlers, there are two components which perform more basic tasks:

- A Metadata Handler, which maintains the metadata catalog and solves the different operations that can be done in the catalog. These operations will be ordered by both the Schema and the Object handlers, and also by the following component.
- A Classical-Fuzzy Translator, whose task is the translation between the events which happen in the fuzzy object-oriented context and the corresponding events that have to be performed on the classical object-oriented context that is located below the Conceptual Fuzziness Handler.

6.4.2   The underlying system

We expect that the underlying database system will be in charge of the main database management responsibilities. The choice of this component of the architecture is not constrained by too many requirements, since the structures proposed to handle fuzzy extensions are built using standard classical object-oriented features.

Nevertheless, we can choose between using a commercial object-oriented database system or using what is called object-relational technology.

This technology lies in the addition of object-oriented concepts to a relational database system, in such a way that the obtained product has many of the advantages of both data paradigms. We can even use a relational database system together with an object-oriented programming language (e.g. JAVA[24] or C++[25]). This alternative is reasonable when we have no object-oriented database system available to implement the fuzzy database system.

6.4.3   FoodBi, a Fuzzy Object Oriented DataBase Interface

In order to experiment with the viability of our proposal, we are currently developing a prototype. FoodBi (Fuzzy object oriented dataBase interface) is a graphical user interface that allows the creation and management of fuzzy object-oriented database schemata. Users can build a hierarchy of classes with fuzzy types where attributes, in turn, can be described using appropriate domains for vagueness handling.

With respect to the architecture depicted in fig. 6.3, FoodBi implements the necessary components to perform the schema management:

- A graphical interface to define schemata, driven by an object-oriented user model.
- The components of the Conceptual Fuzziness Handler required for performing schema management, namely, the Schema Handler, the Metadata Handler, and a partial Classical-Fuzzy Translator.

The FoodBi operation is relatively simple: the user gives the fuzzy data model she/he wants to generate and the application creates the classical object-oriented structures needed to support this model. In particular, our prototype generates a set of classes using the Java programming language.

6.5   Conclusions and Further Work

This chapter intends to be a proposal for representing fuzziness in a classical object oriented database model. Fuzziness may appear at different levels in our database and we have to distinguish between uncertainty and imprecision. Moreover, these two situations may happen at the same time.

We believe that the traditional object oriented database model is wide and general enough to be the basis for developing a Fuzzy Object Oriented Database Model. The structures described to handle fuzzy extensions of the object-oriented data model allow us to develop the architecture of a fuzzy object oriented database system, which can be built upon an existing classical database system.

One of the topics uncovered by this paper is the modeling of the object behaviour when vagueness affects the values of their properties. Though some approaches have been proposed in the literature[26] [27], we think that this matter still needs deeper study, maybe in the context of hypothetical reasoning.

At the same time, we need to study soft ways to retrieve fuzzy information from the database. We are currently working on a generalization of the object comparison operators by means of fuzzy resemblance relationships.

Other part of our research effort is aimed towards the complete development of FoodBi, our prototype, by adding persistence to its fuzzy object-oriented schema definition capabilities.

References

1. J. F. Baldwin, T. H. Cao, T. P. Martin, and J. M. Rossiter. Implementing Fril++ for uncertain object-oriented logic programming. In *Proceedings of the 8th IEEE International Conference on Information Processing and Management of Uncertainty in Knowledge-Based Systems*, pages 496–503, 2000.
2. J. F. Baldwin, T. H. Cao, T. P. Martin, and J. M. Rossiter. Toward soft computing object-oriented logic programming. In *Proceedings of the 9th IEEE International Conference on Fuzzy Systems*, pages 768–773, 2000.
3. M. Berler, J. Eastman, D. Jordan, C. Russell, O. Schadow, T. Stanienda, and F. Velez. *The object data standard: ODMG 3.0.* Morgan Kaufmann Publishers, 2000.
4. E. Bertino and L. Martino. *Object-Oriented Database Systems. Concepts and Architectures.* International Computer Science Series. Addison-Wesley, 1993.
5. G. Bordogna, D. Lucarella, and G. Pasi. A fuzzy object oriented data model. In *Proceedings of FUZZ-IEEE*, pages 313–317, 1994.
6. G. Bordogna and G. Pasi. Management of linguistic qualification of uncertainty in fuzzy databases. In *Proceedings of Information Processing and Management of Uncertainty in Knowledge Based Systems*, 1998.
7. G. Bordogna, G. Pasi, and D. Lucarella. A fuzzy object oriented data model for managing vague and uncertain information. *International Journal of Intelligent Systems*, 1999.

8. T. H. Cao. Uncertain inheritance and recognition as probabilistic default reasoning. *International Journal of Intelligent Systems*, 16:781–803, 2001.

9. R. George, B. P. Buckles, and F. E. Petry. Modelling class hierarchies in the fuzzy object-oriented data model. *Fuzzy Sets and Systems*, 60:259–272, 1993.

10. A. Gonzalez, O. Pons, and M. A. Vila. Dealing with uncertainty and imprecision by means of fuzzy numbers. *International Journal of Intelligent Systems*, 21:233–256, 1999.

11. P. M. D. Grey, K. G. Kulkarni, and N. W. Paton. *Object-Oriented Databases. A Semantic Data Model Approach*. International Series in Computer Science. Prentice Hall, 1992.

12. J. Lee, J.Y. Kuo, and N.L. Xue. A note on current approaches to extending fuzzy logic to object-oriented modeling. *International Journal of Intelligent Systems*, 16(7):807–820, 2001.

13. Z. M. Ma, W. J. Zhang, W. Y. Ma, and G. Q. Chen. Conceptual design of fuzzy object-oriented databases using extended entity-relationship model. *International Journal of Intelligent Systems*, 16:697–711, 2001.

14. N. Marín, I. J. Blanco, O. Pons, and M. A. Vila. Softening the object-oriented database-model: Imprecisin, uncertainty, and fuzzy types. In *Proceedings of IFSA/NAFIPS World Congress*, 2001.

15. N. Marín, M. A. Vila, I. J. Blanco, and O. Pons. Fuzzy types as a new layer on an object oriented database system. In *Proceedings of IPMU*, pages 1099–1106, Madrid, España, 2000.

16. N. Marín, M. A. Vila, and O. Pons. Fuzzy types: A new concept of type for managing vague structures. *International Journal of Intelligent Systems*, 15:1061–1085, 2000.

17. N. Marín, M. A. Vila, and O. Pons. Fuzzy types: Softening structures. In *Proceedings of FUZZ-IEEE*, pages 774–780, San Antonio, Usa, 2000.

18. N. Marín, M. A. Vila, and O. Pons. A strategy for adding fuzzy types to an object-oriented database system. *International Journal of Intelligent Systems*, 2001.

19. S. L. Na and S. Park. A fuzzy association algebra based on fuzzy object oriented data model. In *Proceedings of the 20th International Conference on Compsac*, pages 624–630, 1996.

20. S. L. Na and S. Park. Management of fuzzy objects with fuzzy attribute values in new fuzy object oriented data model. In *Proceedings of the 2nd. International Workshop on FQAS*, pages 19–40, 1996.

21. OMG - Object Management Group. *OMG Unified Modeling Language Specification*. OMG, www.omg.org, march 2000.

22. J-P. Rossazza, D. Dubois, and H. Prade. A hierarchical model of fuzzy classes. In *Fuzzy and Uncertain Object-Oriented Databases. Concepts and Models.*, volume 13, pages 21–61. Advances in Fuzzy Systems- Applications and Theory, 1998.

23. E. H. Ruspini. Imprecision and uncertainty in the entity-relationship model. In H. Prade and C. V. Negiota, editors, *Fuzzy Logic and Knowledge Engineering*, pages 18–28. Verlag TUV Reheiland, 1986.

24. B. Spell. *Professional Java Programming*. Wrox Press, 2000.

25. B. Stroustrup. *The C++ Programming Language*. Addison-Wesley, 1986.

26. N. Van Gyseghem and R. De Caluwe. Fuzzy object-oriented databases: Some behavioral issues. In *Proceedings of EUFIT*, pages 361–364, 1994.

27. N. Van Gyseghem and R. De Caluwe. Imprecision and uncertainty in the ufo database model. *Journal of the American Society for Information Science*, 49:236–252, 1998.

28. R. M. Vanderberghe and R. De Caluwe. An entity-relationship approach to the modeling of vagueness in databases. In *Proceedings of ECSQAU - Symbolic and Quantitative Approaches to Uncertainty*, pages 338–343, 1991.

29. M. A. Vila, J. C. Cubero, J. M. Medina, and O. Pons. *Logic and fuzzy relational databases: A new language and a new definition*, pages 114–138. 1995.

30. M. A. Vila, J. C. Cubero, J. M. Medina, and O. Pons. A conceptual approach for dealing with imprecision and uncertainty in object-based data models. *International Journal of Intelligent Systems*, 11:791–806, 1996.

31. A. Yazici, D. Aksoy, and R. George. The similarity-based fuzzy object-oriented data model. In *Proceedings of IPMU*, volume 3, pages 1177–1182, 1996.

32. A. Yazici, R. George, and D. Aksoy. Design and implementation issues in the fuzzy object-oriented data model. *Journal of Information Sciences*, 108:241–260, 1998.

33. A. Yazici and M. Koyuncu. Fuzzy object-oriented database modeling coupled with fuzzy logic. *Fuzzy Sets and Systems*, 89:1–26, 1997.

34. L. A. Zadeh. Fuzzy sets. *Information and Control*, 8:338–353, 1965.

35. A. Zivieli and P. P. Chen. Entity-relationship modeling and fuzzy databases. In *Proceedings of the Second International Conference on Data Engineering - IEEE*, pages 18–28, 1986.

7 Domain Analysis for the Engineering of Spatiotemporal Software

Ali Dogru and Adnan Yazici

Computer Engineering Department
Middle East Technical University
Ankara, Turkey
email:dogru@ceng.metu.edu.tr , yazici@ceng.metu.edu.tr
(90-312)210-5590
(90-312)210-1259

7.1 Introduction

Software field in the expected wake of the industry from the long suffered "software crisis," is implementing the approach discovered long time ago in "hard" engineering disciplines. Well-defined blocks of smaller solutions that can be used in the synthesis of any system have been the key behind fast and reliable design. For any engineering field there are widely recognized sets of building blocks that facilitate development. The alternative is to build complex systems starting with the minutest details; hence reinventing the wheel each time.

The technology is called components. Recently component protocols that define the architecture and the integration rules for the components have emerged along with component sets to serve in different software engineering domains. Utilizing this technology, a new paradigm is in formation, that is "Build by Integration" rather than code writing [19]. Justification for the trend is not difficult: it is not practical to develop millions of lines of code in a timely and reliable manner. The demand for ever-increasing complexity suggests that a linear effort in the composition of huge systems will not just be sufficient. There will be need for more serious treatment of the interface concept [18]. There could also be hardware [4] as well as software modules to add to the complexity.

7.1.1 Domains

Identifying, locating and integrating components, however, are only feasible in a defined domain. It is also not practical to know every component in all fields intimately. The future success of the technology in software engineering will depend on the maturity of the domains for which, standardization and the recognition of the components by their names will be a defining factor [17]. Features, processes, and terminology that are meaningful for a domain will help in the identification of the components.

There have been some success reports in the practice of newly developing Domain Oriented Software Development field [2,15,11–14,16]. Not only implemented building blocks, but also design patterns [8], architectural structures, and knowledge at a variety of levels are classified and made available for access for the development process. Frameworks are versatile tools for utilizing these concepts [7]. The approach facilitates reuse more than in the coding level. In this chapter special kinds of problem features are addressed in the domain analysis perspective. These are the uncertain, spatial and temporal issues that could add to the complexity of a variety of problem fields. Actually, more than one domains relate to these features and can incorporate the suggested infrastructure.

7.1.2   The Spatiotemporal Domain

Software systems are categorized with respect to different parameters. A frequently utilized parameter is the timeliness of the systems response. These parameters, however, relate to the run-time behavior, or the complexity of the written code. Domains, on the other hand, provide a different categorization criterion that is more related to the problem definition rather than its solution. A family of problems and their solutions can define a domain.

It is possible to further categorize the domains for software development, with respect to a horizontal and vertical partitioning of the universal application fields. A hierarchical organization will help in the categorization of domains. As in the general categorization problem, our effort is not free from the difficulties such as a domain's taking place in different categories organized with respect to different criteria.

The spatiotemporal super-domain is actually a category of real domains. Geographical information systems domain, Command Control and Communications applications domain, and Medical Imaging domain are examples of spatiotemporal domains. The problems dealt in those domains have to address both spatial and temporal phenomena. Although not widely modeled to reflect, they also have fuzzy characteristics. The variations in time or space can be represented in fuzzy parameters very conveniently.

7.2   Domain Analysis

Domain analysis is a key activity in the Domain Oriented Software Development paradigm. The new trend is towards building software by integration, rather than writing code. Component technologies are the enabling factor for the process. Components, as implemented pieces of code will be retrieved from the domain model and integrated to yield a specific application. Fig. 7.1 depicts the utilizations of components in a domain.

Domain Analysis activity is the key effort in the populating of a domain. What is required is the experience from the domain experts, and a study

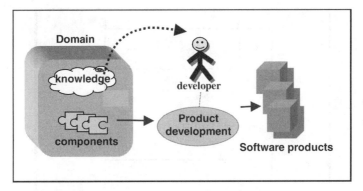

Fig. 7.1. Domain Oriented Software Development

of the existing applications defining the domain. The repeatable features for to-be-developed applications need to be extracted from the domain, documented, and modeled. A domain typically contains a terminology library, a component library, some architectural support and various representations for any domain experience that could be documented.

7.2.1   Primitives for the Semantic Modeling

The uncertainties inherent in spatiotemporal problems have to be represented. One such study has been conducted [21] that yielded a Unified Modeling Language (UML) based representation [3]. Class and association types were extended in UML to achieve the required representations.

The three types of uncertainties addressed are:

1. Incompleteness: True data belongs to a set of specific values,
2. Null: True data value is not known, and
3. Fuzzy: True data is available but not expressed precisely.

The uncertainty types are represented by "Ut_{in}, Ut_{nu}, and Ut_{fy}," for incompleteness, null, and fuzzy, respectively. A class containing uncertain attributes is treated as a special class and it is extended from the UML class. To denote the existence of uncertainty in a class, a "U" symbol is placed to the left of the class name, as shown in Fig. 7.2.

The uncertainty mentioned so far concerns the attributes in a class. For a class itself representing a fuzzy membership of its instances to the corresponding object set, however, a different symbol is used. The "U" symbol is placed in a double square as shown in Fig. 7.3, and the membership degrees of its instances will be indicated by a real number (0..1) assigned to the "membership" attribute.

Also, temporal objects should be accounted for. Some or all of the attributes of an object may change with time. Such objects are viewed as "temporal" and their class definitions must be associated with a "time" object. A

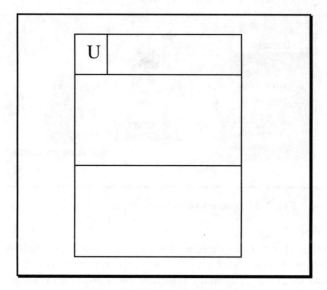

Fig. 7.2. The uncertain class

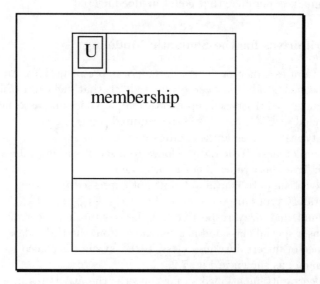

Fig. 7.3. Class for fuzzy membership

time object is represented by placing the symbol "T" to the right of the class name as shown in Fig. 7.4.

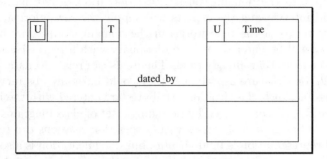

Fig. 7.4. Fuzzy temporal objects

7.3   Domain Model

Typically, after populating the domain with knowledge, it will be reused several times for generating new applications. Actually, the domain model is a dynamic repository; as new information develops, the model should be updated. There are various approaches to structuring, populating, and utilization of a domain [12]. There exist methodologies for the purpose [11,13,1]. This chapter does not introduce a detailed methodology. The objective here is to highlight the characteristics of spatiotemporal domains in general and a possible way to achieve a domain oriented methodology for the purpose.

The proposed domain model contains four types of repositories:

1. Domain Terminology Dictionary,
2. Feature Base,
3. Architectural Patterns, and
4. Component Framework.

The Dictionary is similar to a "Data Dictionary" introduced in the early era of the software engineering. Any term is entered with its explanation in relation with other terms. The Feature Base is the heart of the model, where classes and associations represent the structural and conceptual components defining the field. Architectural patterns provide a template structure for parts of the system under development. A skeleton is provided which will render executable code, when fleshed with components. Architecture provides the connectivity together with the structural foundations for the products. The executable pieces to adhere to and make architectural patterns run, are the components. They are retrieved form the Component Framework that offers a library of components and necessary services for their adaptation and wiring to the architecture.

7.3.1 Developing a Domain Model

Documents, existing applications, and the knowledge of the domain experts are the input to the domain constructing operation. Analyzing the applications and interviewing the experts will supply the raw information to be modeled in terms of the repositories in the domain model. Any important terminology will be entered into the dictionary with a regressive update to the related entries that already exist. The items can relate to data, function, control [22], or structure aspects and they could have any abstraction level. Where possible, such classification attributes are entered with the item.

Feature Base is constructed by providing a set of class diagrams. Isolated classes as well as groups of classes with association relations can take place. Graphical representation is fundamental, but documentation is also very important. Besides known formal techniques for the refined specification of the classes, expertise can be explained in terms of suggested usage, examples of adaptation, special cases and the like.

Architectural patterns represent a repeated usage of similar program structures. A set of similar components or objects can be re-implemented in a number of applications with a similar connection topology. These common patterns can be provided as larger-grained components to be located and used after modification. When completely instantiated, they would constitute relatively large parts of the applications. Searching for repeated architectural chunks in existing applications is the key task for identifying patterns and introducing them to the domain model.

Components are the most practical constituents of a domain model. They are built for integration by the third parties. They are well defined with interfaces to define their services. A component is not built for a specific system - it is supposed to be for general use. A set of components is required to be developed. A careful analysis of the functionalities and how they could be organized into components is the beginning step for the construction of the framework. General modularity concerns guide the determination task for the functionality scope of each component. Next comes the determination of the interfaces of a component.

Besides the default requirements to satisfy the underlying component protocol, meaningful organization of interfaces has to be achieved. An interface usually groups closely related services, especially considering what kind of 'clients' could request. The interfaces have to be documented in detail since they are the specification for the component. Its implementation could change in time, as a modification to the domain model. The new implementation should be as faithful as possible to the previously declared interfaces.

If the applications will be generated obeying one component protocol, the component framework with the implementations of the components and the wiring aid, should completely support the protocol. If alternative protocols are supported, different versions of the components should be developed whenever feasible. Component adaptors or bridges can be used but it is better to have the whole component to be restructured with respect to the new component. It is possible, however, to provide a framework without an established protocol. The home-grown protocol should be the substitution, at least enforcing a well-defined interface definition. The two relevant prominent protocols are DCOM and JavaBeans. There has also been attempts to offer component based methodologies to address the software lifecycle [9,10,20].

7.3.2   Utilizing a Domain Model

Now that the infrastructure is ready, the process for software development can begin. The process is heavily based on reuse practices though. Requirements of the specific application will be documented as expected. Then, an earlier start of design related activities takes place. Requirements tasks in part are already achieved in the domain model. Only the specific requirements that exceed the commonalities of the domain need to be investigated in detail. Tools to conduct a "what if analysis" at the requirements level [6] may guide a significant contribution here. From the outset, implementation alternatives will be considered and appropriate architectural patterns will be identified. For the detailed design, component adaptation and connection peculiarities will be resolved. As components are integrated, tests should be carried out before the system is composed entirely. Finally, evolutionary paradigms suggest to be ready to revisit the earlier decisions: some component adaptation, selection, and even architectural decisions could be proven inappropriate at the later stages.

7.3.3   The Radar Signal Monitoring Domain

This section introduces an example domain for spatiotemporal applications. Radar signal monitoring is selected due to the inherent spatial and temporal values with uncertainties involved. When used for detecting the aircraft, radars produce track information. Actually, the current three-dimensional coordinates of the detected object are being reported repetitively in a fraction of a second. The control software should be able to relate the reported coordinate to the previous ones, hence determining the flight paths. There could be about fifty traces on the screen, all needing an update with the new scan of the monitored space.

1. Uncertainties
   The newly reported coordinates could be the decided to be an extension of one of the traces, but this information is not very crisp. Also, traces may disappear falling behind a hill or some other obstacle. When they reappear from another location, they should be connected with one of the previously tracked traces. Sometimes two traces may intersect at close angels and it may be difficult to identify them after the intersection. Identification of an aircraft as friend or foe is also an issue and not always

easy. The regions are important, posing the spatial problem. How close is an object to the surface under defense is important information. The geographical model of the area should be integrated within the application.

2. Modeling the Example Domain

Selected items will be included in this section, as demonstrative elements of the model. Analysis of a set of existing applications, interviews with the domain experts, and studying of the related documentation/books are assumed done. While features, patterns, and components are being incorporated, every related terminology is entered into the dictionary. Table 7.1 depicts the Domain Terminology Dictionary for the selected entries.

Table 7.1. Domain Terminology Dictionary

Trace			
	Explanation:	A string of points. Corresponds to a flying object.	
	Attributes:	FoF:	friend/foe/unknown
		Threat:	none/low/medium/high/urgent
		Speed:	real number
		Course:	current tangent to the trace
	Relations:	Point:	trace=string of points
		Region:	proximity, targeting
Point			
	Explanation:	A vector of 3 real numbers representing Cartesian coordinates.	
	Attributes:	Membership:	The degree of its membership to a trace
		Azimuth:	Angle of height
	Relations:	Trace:	A string of points=trace
		Region:	proximity

The features of the domain are related with the traces and temporal and spatial entities. Fig. 7.5 depicts the UML class diagram for the spatiotemporal properties of the region a radar system is supposed to monitor for defense. The center of the diagram is a spatial object (Spatial_O) class. This entity generalizes a regional feature such as a city, a forest etc. Many spatial objects can be in an association to represent a topology. Also a spatial object is dated by one or more time objects which mark time-windows for enemy threat. The spatial object class can have many geometry objects. A geometry object is identified by one of the geometry primitives, namely:

- a three dimensional object,

- a two dimensional object,

- a line, or

- a point.

Actually, Geometry is a feature of the Position class.

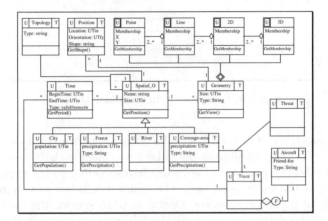

Fig. 7.5. Spatiotemporal features in the region covered by radar

In some cases, the compression of the geometry is desired. A three-dimensional geometry can reduce to two dimensions as a result of compression. Similarly, a two-dimensional geometry can reduce to a line and a line can be compressed to a point. Fig. 7.5 does not reflect every class and does not represent every relation among the displayed classes. A more detailed view for the geometry features is presented in Fig. 7.6. If there is compression, the higher-level geometry object will render a single instance of the lower-level geometry object. Therefore the compression relation is an association with the pluralities of "one to zero" or "one to one." On the composition side, the relation requires a higher-level geometry and at least two lower-level geometry objects.

Before introducing architectural patterns, an example set of components will be presented. The patterns may actually be defined in terms of components.

3. Components

First component introduced in this section is the radar. This piece of software is capable of communicating with hardware and maintaining the traces detected. A graphical display is derived and interactive input is accepted. Abilities include the tagging of the traces as friend/foe, and with the threat degree. The algorithm for correlating a new point with one of the existing traces can be selected. Geographical regions can be entered for the automatic generation of threat events when a foe trace approaches them. Also a cancel threat event is produced. Common component protocols allow for subscribing to event types and responding to

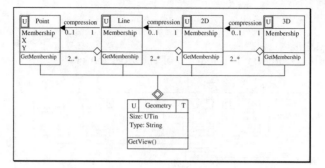

Fig. 7.6. Compression in Geometry Objects

the events when they notify the subscriber component with the occurring of the event. Fig. 7.7 depicts the radar component with its two interfaces. The hardware interface (i_hardware) is responding to the requests originating at the radar hardware. The monitor interface (i_monitor) is for serving the operator requests. Drawn in UML syntax, the radar component will be detailed in a COSEML [5] representation for the interface related refinement in Fig. 7.8.

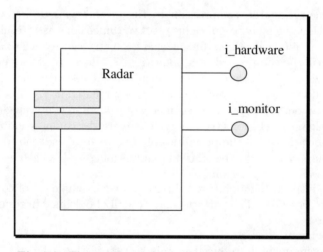

Fig. 7.7. The radar component in UML representation

Most protocols require that the events listed in the interface of a component should be in two separate categories; for input and for output. Actually, components are being viewed as mostly "server" kind of objects. The services they provide take place also in the interface as methods. Outgoing message types are usually not modeled. Actually, for the integration of components, outgoing messages are as important as the incoming messages (methods in the interface) [5]. Like in the cases of events, methods can be specified in two groups; incoming service requests by other components, and outgoing requests required by the current component.

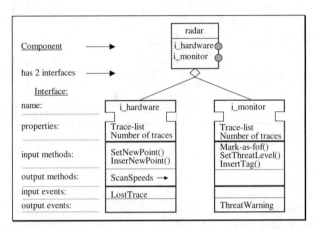

Fig. 7.8. The interfaces for the radar component in COSEML representation

An example pattern constitutes the radar, radar hardware, aircraft, console, and geography components. Fig. 7.9 displays the pattern with its dependencies in UML syntax. The dependencies correspond to connections among the components. Virtual connectors represent message traffic. The pattern can be detailed by collaboration diagrams to reflect the message sequences in the realization of the domain related scenarios. Some of the messages and even connections may be unnecessary for the current application and do not have to be instantiated. Nevertheless, the pattern serves as a template to offer a sub-solution after possible modifications. In a more general use the components could be substituted and the skeleton pattern with connectors will be the contribution by the pattern. The patterns can be hierarchically organized but this is not recommended for most of the cases. There will be patterns for different levels of granularity anyway. The container for all the patterns is the architecture, hopefully implemented by a domain tool supporting a framework kind of environment.

4. Utilizing the Domain

In a typical domain model, there will be many components and many patterns. Some of the items will be repeated with modifications for alternative selection. A component may provide functionality with different algorithms, selectable during the adaptation or modification process. Some algorithms may be more involved requiring to be encapsulated in specific components. Therefore, a similar functionality could be formulated by different algorithms and implemented by different components.

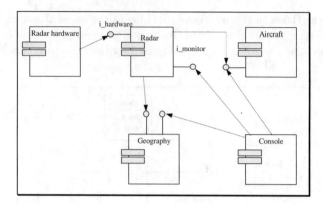

Fig. 7.9. An example pattern for the radar signals monitoring domain

Also architectural patterns could offer alternatives. Sometimes, the alternative structures may have overlapping sections. A single pattern may be sufficient for some rare applications, but most of the time, different patterns will be integrated to construct an application.

Pattern or component, the items in the domain are ready for browsing, reusing, and modifying for a composition. A methodological approach would suggest the envisioning of the system as a whole and decomposing it into its abstract modules. Depending on their granularity levels, these modules will correspond to patterns or components. The unit that satisfies abstract modules specifications is the initial candidate for utilization. If the item matches the specification almost exactly, it can be reused as it is. For less perfect matches, some modification will be necessary. For the optimistic cases, these modifications will correspond to the supported adaptation procedure for the component. Adaptations and modifications indicating a frequent usage indicate the need for an update for the domain model. Finally by integrating the items fetched from the domain model, the product can be generated. Integration will most of the time mean notifying a component with the identity of another, and the identity of another components interface.

Additional code may be required. Care must be taken for remaining within a protocol and within component-only style. Frequent usage of a pair of components should start the investigation on composing those two components into a third bigger component and offering the initial two and the produced third in the domain. A composition means some code writing to produce the new interface(s) for the new component. In some cases, the composed component declares the interfaces of the contained components as its own, hence reducing the need for new code writing.

7.4   Conclusions

A domain-oriented approach for the spatiotemporal software engineering problems is introduced. The complex features of such class of problems require a structured organization of special features. In the suggested domain modeling, a class diagram based representation of features is employed, supported by a domain terminology dictionary. For the implementation stage support, the domain offers a set of components and architectural patterns for composing an application.

The example domain and its model is presented for the demonstration of the spatiotemporal problem representation; actual implementation of such a model has not been carried out. Although similar domain work have been conducted without a domain-orientation, and spatiotemporal modeling and development were experienced before, the outcome has yet to be utilized in the proposed context.

It can be observed that peculiarities pertaining to spatiotemporal fields can yield more efficient engineering if structurally modeled in a domain environment. Besides the intellectual control over the field it provides, the domain environment will definitely help in the development of applications through its powerful reuse practices. Domain-based development has been an effective way for software development. The leverage is especially important when used for the application fields that accommodate especially complex features. It is hoped that the future work will result in practical domain environments for real-world applications. After then, the expected verification of the idea can be discussed.

References

1. Ilkay Altintas. *A Comparative Study for Component-Oriented Design Modeling*. M.S. Thesis, Computer Engineering Department, Middle East Technical University, Ankara, Turkey, May 2001.
2. G. Arrango. Domain analysis methods. In W. Shcaefer, R. Prieto-Diaz, and M. Matsumoto, editors, *Software Reusability*. Ellis Horwood, 1994.
3. Grady Booch, James Rumbaugh, and Ivar Jacobson. *The Unified Modeling Language User Guide*. Addison-Wesley, 1999.
4. A.H. Dogru, S.N. Delcambre, C. Bayrak, Y.T. Chen, E.S. Chan, W. Yin, M.G. Christiansen, and M.M. Tanik. An integrated system design environment: Concepts and a status report. *Journal of Systems Integration*, 2(4):317–347, October 1992.
5. Ali. H. Dogru and Ilkay. Altintas. Modeling language for component-oriented software engineering: COSEML. In *The Fifth World Conference on Integrated Design and Process Technology*, Dallas, Texas, June 4-8 2000.
6. Huseyin Dursun and Ali H. Dogru. Prototyping specifications through visualization. In *The First World Conference on Integrated Design and Process Technology*, volume 1, pages 362–368, Austin, Texas, December 8-9 1995.

7. Mohamed E. Fayad. Introduction to the computing surveys' electronic symposium on object-oriented application frameworks. *ACM Computing Surveys*, 32(1):1–11, March 2000.

8. Eric Gamma, Richard Helm, Ralph Johnson, and John Vlissides. *Design Patterns: Elements of Reusable Object-Oriented Software*. Addison Wesley, Reading, Massachusetts, 1995.

9. George T. Heineman and William T. Councill. *Component-Based Software Engineering*. Addison Wesley, 2001.

10. Peter Herzum and Oliver Sims. *Business Component Factory*. Wiley, 2000.

11. Robert Holibaugh. Joint Integrated Avionics Working Group (JIAWG) Object-Oriented Domain Analysis Method (JODA), CMU/SEI-92-SR-3. Technical report, Software Engineering Institute, Carnegie Mellon University, Pittsburgh, Philadelphia, November 1993.

12. Kiyoshi Itoh, Toyohiko Hirota, Satoshi Kumagai, and Hiroyuki Yoshida (editors). Domain oriented systems development: Principles and approaches. *Information Processing Society of Japan*, 2000.

13. Kyo C. Kang, Sholom C. Cohen, James A. Hess, William E. Novak, and A. Spencer Peterson. Feature-oriented domain analysis (FODA) feasibility study, CMU/SEI-90-TR-21. Technical report, Software Engineering Institute, Carnegie Mellon University, ADA 235785, Pittsburgh, Philadelphia, 1990.

14. J.M. Neighbors. DRACO: A method for engineering reusable software systems. In *Software Reusability*, volume 1, pages 295–320. ACM, 1989.

15. Ruben and Prieto-Diaz. Domain analysis for reusability. In *COMPSAC 87: The Eleventh Annual Computer Software and Applications Conference*, pages 23–29, October 1987.

16. M. Simos. Organization domain modeling (ODM): Extending systematic domain analysis and modeling beyond software domain. In *IDPT 1996*, 1996.

17. Clemens Szyperski. *Component Software: Beyond Object-Oriented Programming*. Addison Wesley, New York, 1998.

18. M.M. Tanik and A. Ertas. Design as a basis for unification: System interface engineering. *ASME PD*, 43:113–114, 1992.

19. M.M. Tanik and A. Ertas. Interdisciplinary design and process science: A discourse on scientific method for the integration age. *Journal of integrated Design and Process Science*, 1(1):76–94, September 1997.

20. Kurt C. Wallnau, Scott A. Hissam, and Robert C. Seacord. *Building Systems from Commercial Components*. Addison Wesley, 2002.

21. A. Yazici, Q. Zhu, and N. Sun. Semantic data modeling of spatiotemporal database applications. *International Journal of Intelligent Systems*, 16:881–902, 2001.

22. Weiping Yin. *An Integrated Software Design Paradigm*. Ph.D. Dissertation, Southern Methodist University, Dallas, Texas, 1988.

8 Object-Oriented Framework of Fuzzy Knowledge Systems

Mong-Fong Horng, Shiuh-Chu Lee and Yau-Hwang Kuo

Center for Researches of E-life Digital Technologies (CREDIT)
Department of Computer Science and Information Engineering
National Cheng Kung University, Tainan, Taiwan

8.1 Introduction

In the real world, many sophisticated AI problems are decentralized, vague, and cooperative in the issues of problem solving and knowledge manipulation. Therefore, we need an effective framework suitable for modeling the problem-solving activities realized by multiple actors, which may employ various reasoning schemes according to different conditions, with vague domain knowledge and communication manner. Obviously, object-oriented paradigm [2] is an adequate solution due to the characteristics of encapsulation, modularization, polymorphism, inheritance, and message passing. In fact, in recent years, object-oriented modeling approach has been applied to AI topics [5,1,8,3,7,4].

However, there are several issues that should be clarified to model sophisticated AI problems by the object-oriented paradigm. How to embed approximate reasoning capability into the proposed object-oriented framework is the first issue that should be considered. Fuzzy set theory is obviously a good solution for this issue [9]. An object-oriented fuzzy system model can represent and manipulate vague problem-solving knowledge in a structural manner.

Where the inference mechanism locates is the second issue. Some different schemes may be applied. For instance, the inference engine may be implemented as a stand-alone agent to operate on the other objects which constitute the static knowledge base. In such case, object orientation only happens on the knowledge base. For another extreme case, every object in the modeled system is a coarse-grained object which contains a traditional inference engine and a large part of domain knowledge. This case is suitable for the cooperative problem-solving systems running in a loosely-coupled distributed computing environment. Besides, there should be the third case for AI or expert systems which are adequate for running in a tightly-coupled parallel processing system. In this case, multi-threaded reasoning capability with maximum parallelism and fine-grained object are expected. Any modeling framework may be suitable for one or more of the three cases.

The third issue that should be clarified is how to realize the inheritance process in a vague situation. Conventional object-oriented paradigm requires crisp inheritance, but in the real-world inheritance may be disturbed by some factors. Therefore, an inheritance mechanism with partial inheritance capability is necessary.

In this article, we develop an object-oriented fuzzy knowledge system framework that can solve the issues mentioned above. In the proposed framework, Fuzzy Linguistic Objects (FLOs) are the basic constituents, and each of them represents a group of related fuzzy linguistic variables. The properties of object-oriented paradigm such as encapsulation, inheritance, polymorphism, and message passing are all employed in the proposed modeling framework. By this way, knowledge base is realized by corpus of linguistic objects, and the inference process is constituted by interactions (message passing) and inheritances among objects. Therefore, the proposed modeling framework has higher potential to exploit the parallelism in reasoning and manipulate problem-solving knowledge in a distributed environment.

8.2   Fuzzy Linguistic Objects (FLOs)

The notion of FLO is extended from the definition of fuzzy linguistic variable.

8.2.1   Object orientation of linguistic variable and knowledge representation

A linguistic variable is a higher type fuzzy set using linguistic terms as values [10]. Obviously, the notion of linguistic variable may be hierarchical but not structural. In fact, a linguistic variable can be viewed as an attribute of an object. Such an object is called **Fuzzy Linguistic Object (FLO)**, which is defined as $FLO = (i, A, T, F, C, P)$, where i is the identity of the FLO, A is the linguistic attribute set whose elements A_j's are linguistic variables, is the linguistic term set, F is a set of mappings where each element $F_j : A_j \rightarrow T$ denotes the linguistic terms associated with linguistic attribute A_j, C is the component object set, and P is the parent object set of the FLO. In other words, C presents the aggregation (has-a) hierarchy, while P denotes the generalization (is-a) hierarchy. Based on the FLO paradigm, a fuzzy knowledge system can have richer knowledge expression power than traditional rule-based scheme. Domain knowledge in the FLO-based framework is represented by three parts: generalization hierarchy of FLOs, aggregation hierarchy of FLOs, and object-oriented fuzzy rules. The former two parts represents the structural knowledge while the last one represents heuristic rules for problem-solving. The format of rules is "if FLO_i.attributej is A then FLO_k.attribute$_m$ is B" where A and B are linguistic values of attributes i and j respectively, and the premise part can be a compound condition.

8.2.2   Organization of fuzzy linguistic object

The structure of a FLO is shown in Fig. 8.1, which includes (1) message exchange interface: the query processor, reference set, and binding set, (2) reasoning mechanism: the matcher set and conflict resolver and (3) object information unit: the attributes and component set.

(A) The matcher set. Every matcher is an object that deals with the transit of information [6]. A fuzzy matcher is a special kind of matcher that has the function to execute fuzzy inference of generalized modus ponens. It encapsulates a linguistic term A in the form of membership function and two functional units: **matching degree calculator** and **implication calculator**, as shown in Fig. 8.2. The matching degree calculator is activated when the linguistic term A appears in the premise statement of the rule being fired. It accepts a fuzzy input A', gets the membership function of A, calculates the matching degree between A and A' with a chosen composition operation, and finally sends the matching degree out. This process is called "match out." On the other hand, the implication calculator is activated when the term A is expressed in the conclusion part of the rule being fired. It accepts a matching degree σ from another FLO, and then calculates a partial fuzzy conclusion A" for A according to a certain fuzzy implication scheme. This process is called "match in." By the polymorphism property of object-oriented paradigm, different types of matching degree calculators (implication calculators) can be implemented for different composition operations (implications), but all of them have the same invocation interface. Therefore, more than one fuzzy inference scheme or fuzzy operation type can be included into the same FLO model, which provides more flexibility to schedule the inference behavior.

(B) The reference set. A reference unit, as shown in Fig. 8.3, is constructed when a request for establishing an inference channel between two FLOs is allowed. In the conceptual level, a reference is a symbol to represent a fuzzy concept associated with one attribute of a FLO. The target pointer in Fig. 8.3 records the address of the FLO to which the information will be sent. For instance, assume a reference of FLO, O_1, presents the premise part of the fuzzy rule "if O_1.x is A then O_2.y is B," then it will contain a fuzzy matcher to represent the fuzzy concept A. During reasoning, this reference gets the value of x, A', from the attribute set of O_1, calculates the matching degree between A and A', and finally sends the matching degree to FLO O_2 according to the address specified in the target pointer.

(C) The binding set. A binding is also a symbol that represents a fuzzy concept associated with one attribute of a FLO. In the FLO framework, an inference channel is constructed by a (reference, binding) pair which will be discussed in Section 8.4, and the direction of information flow is from the reference to the binding. The reference is constructed to supply necessary information for inference, while the binding is constructed to make inference. As shown in Fig. 8.4, a basic binding element contains a source pointer and a binder. The former stores the address of the object from which the informa-

tion is gotten (that is, the location of the corresponding reference in another FLO), and the latter is an information requesting mechanism. For executing fuzzy reasoning, there are two kinds of bindings in the binding Set: the evidence bindings and the inheritance bindings, as shown in Fig. 8.5. Inheritance bindings deal with partial inheritance which will be discussed in next section. It gets the value A of an attribute from the parent object, then calculates the result A' based on an inhibiting parameter and an enforcing parameter by invoking the inheritance processor to work. An evidence binding handles a fuzzy implication relation (fuzzy rule); it gets the matching degrees of premises and calculates a partial conclusion, then sends the partial conclusion to Conflict resolver. Conflict resolver gathers all partial conclusions related to the same fuzzy attribute to generate a global conclusion. In an evidence binding, the evidence set contains a set of basic binding elements to gather the matching degrees of all specified premises. For example, to handle the fuzzy implication, "if O_1.w is A and O_2.x is B and O_3.y is C and O_4.z is D then O.s is E," the evidence set must contain four basic binding elements to manipulate the four premises respectively. Evidence combiner is responsible for the evaluation of total matching degree (rule firing strength) from the gathered partial matching degrees. The fuzzy logical AND operation is the most popular strategy to realize evidence combination. By object-oriented paradigm, Evidence combiner can encapsulate different evidence combination schemes but have uniform interface.

(D) The query processor. This unit deals with the queries for establishing inference channels from the other FLOs. If a query is accepted, it constructs a reference unit with a proper matcher and responses to the querying FLO. Then, the two FLOs establish an inference channel.

(E) The conflict resolver. The structure of conflict resolver is shown in Fig. 8.6, which contains a dispatcher and some conflict resolution units. Each conflict resolution unit is responsible for the evaluation of the final reasoning result (from inheritances and implications) of an attribute in the FLO. The number of conflict resolution units is equal to the number of attributes in a FLO. Dispatcher dispatches information from bindings to corresponding conflict resolution unit.

(F) The attributes. The value in each attribute unit is updated when a new conclusion about itself arrives from the conflict resolver, and can be retrieved by the references.

(G) The component set. Component set is the collection of symbols that stand for the component FLOs. The inference channels between a FLO and its component objects are also constituted by (reference, binding) pairs.

8.3   Partial Inheritance Model

Inheritance provides a method of reasoning with implicit facts [6]. Partial inheritance is defined as a case that an object inherits a property partially from its superclass object. This phenomenon frequently exists in the real world. The idea of the proposed partial inheritance model is that the inheritance of properties from an object to its child object may result in some evolution due to circumstance factors. The evolution may occur in inhibiting or enforcing direction whose strength is represented by inhibiting parameter and enforcing parameter respectively. When the inhibiting effect is stronger than the enforcing effect, the inherited property will be inhibited; otherwise that property will be enforced. Complete inheritance is a special case of partial inheritance that occurs when the inhibiting effect is equal to the enforcing effect.

Suppose (1) O_1 and O_2 are FLOs and x is their one attribute, (2) $O_1 \rightarrow O_2$ means that O_2 inherits from O_1, (3) $value(x)$ is denoted as the linguistic value of x, and (4) • is an aggregation operator defined for partial inheritance. Then, for each x, $\{x | x \in O_1 \, and \, x \in O_2, O_1 \rightarrow O_2\}$, there exists an inheritance relation I_x such that $value(O_2.x) = value(O_1.x) \bullet I_x$ and $I_x = (P_I, P_E)$ where P_I is the inhibiting parameter and P_E is the enforcing parameter. P_I and P_E could be fuzzy sets or crisp values. Furthermore, the aggregator is formulated as follows:

$$value(O_2.x) = value(O_1.x)(P_I, P_E)$$
$$= \cap_U min\{1, \mu_{value(O_1.x)}(u) - \mu_{value(O_1.x)}(u) \mu_{PI}(u) \mu_{PE}(u)\}/u \tag{8.1}$$

where U is the universe of discourse of x, $\mu_{value(O_1.x)}$ is the membership function of linguistic term $value(O_1.x)$.

For the phenomenon of multiple inheritance, suppose an object O_1 inherits the attribute x from the other objects O_i, for i=1, ...,n, then

$$value(O_i.x) = \begin{cases} value(O_2.x) \bullet I_{x2} \\ value(O_3.x) \bullet I_{x3} \\ \dots \\ value(O_n.x) \bullet I_{xn} \end{cases} \tag{8.2}$$

Therefore, we must deal with the conflict phenomenon due to multiple inheritance for the decision of the proper result of value(O_1,x). In FLO framework, the form of multiple rule firings is

$$value(O_1.x) = \begin{cases} value(O_{r1}.y_1)oR_1 \\ value(O_{r2}.y_2)oR_2 \\ \dots \\ value(O_{rn}.y_m)oR_m \end{cases} \tag{8.3}$$

where y_i is an attribute of object O_{ri}, $and R_i$ is the fuzzy implication relation expressing a fuzzy rule " if $O_{ri}.y_i$ is A_i then $O_1.x$ is B_i," for i=1, ..., m. Obviously, the partial inheritance relation could be viewed as a special type of fuzzy rule, and the multiple partial inheritance problem is the same as the conflict resolution problem of fuzzy inference.

8.4   Reasoning in FLO-based Framework

In the FLO framework, the problem solving activity is accomplished by the interactions of FLOs. Basically these interactions can be distinguished into three different inference types: implication, aggregation, and inheritance. Every interaction of any inference type is realized by an inference channel maintained by a (reference, binding) pair, as shown in Fig. 8.7. For the former two inference types, evidence bindings are used, while the last inference type adopts inheritance bindings.

The (reference, binding) channels are dynamically established during problem solving. In the initial state, there is no connection among FLOs. When an interaction for reasoning between two FLOs is initiated but the connection has not yet been established, a binding in the conclusion FLO will be constructed and then sends a message to the query processor of the premise FLO for establishing an inference channel. If the premise FLO allows this asking, it will construct a reference with a compatible fuzzy matcher and return the address of the reference to the binding of the conclusion FLO, then an inference channel between these two FLOs is established. Once an inference channel is established, it is maintained until the system solves the given problem.

The three inference types have some differences in the linking of two FLOs for reasoning. The implication needs a searcher to find the queried FLO. The aggregation does not require a searcher, because the locations of component objects have been recorded during initial construction. The inheritance is similar to the aggregation.

The implication between FLOs may have compound premise. In the proposed framework, the compound premise with AND connectives can be manipulated by the evidence binding, where its evidence combination mechanism executes the AND operation to produce a total matching degree then the fuzzy matcher produces a partial conclusion for the implication. The compound premise with OR connectives are decomposed into simple implication relations that do not contain OR connective. Then, the conflict resolver is responsible for the composition of partial conclusions of all decomposed simple implications sent from the corresponding bindings.

In the FLO framework, the direction of inference can be forward chaining, backward chaining, or mix of both. The direction of inference depends on the behavior of the (binding, reference) pairs. If the reference sends information to binding automatically then the inference is forward chaining. On the other hand, if the reference sends information as the response of a request from the binding, then the inference is backward chaining. Of course, backward chaining and forward chaining can be mixed because the encapsulation and polymorphism properties of FLOs. The chaining scheme is decided when a model is constructed.

Finally, we illustrate the FLO framework by the following example. Suppose (1) O_1,..., O_6 are FLOs where x and s are attributes of O_1, y is an attribute of O_2, t is an attribute of O_3, and z is an attribute of O_4 and O_5; (2) O_1 is a component object of O_6; (3) A1 and A2 are linguistic terms of x, C is linguistic term of s, B_1 and B_2 are linguistic terms of y, D is linguistic term of t, and E1 and E2 are linguistic terms of z (these linguistic terms are encapsulated as fuzzy matchers, $M_{A1}, M_{A2}, M_C, M_{B1}, M_{B2}, M_D, M_{E1}$ and M_{E2} in the corresponding FLOs); (4) the relationships among the FLO objects are

R_1="if O_1.x is A_1 then O_2.y is B_1,"
R_2="if O_1.x is A_2 then O_2.y is B_2,"
R_3="if O_1.s is C then O_3.t is D,"
R_4="O_5.z inherits from O_4,"
R_5="if O_2.y is B_1 and O_3.t is D then O_5.z is E_1."

Then, a FLO model which is briefly shown in Fig. 8.8 can be achieved. Now, suppose value(O_1.x)=A', value(O_1.s)=C' and value(O_4.z)=E_2, and we want to get value(O_5.z). The inference process is shown in Fig. 8.9. Object O_2 receives two pieces of information (matching degrees) from $O_6.O_1$.x and produces two partial conclusions, B'_1 and B'_2. These two partial conclusions are combined by a conflict resolver to produce the conclusion B' to be stored in the attribute y. The conclusion of attribute t, D', in O_3 is gotten in the same manner. On the other hand, object O_5 inherits attribute z from O_4. Although the original value of z is E_2, it becomes E_2' after passing through an inheritance binding due to partial inheritance. Finally, O_5 produces the conclusion of attribute z, E', by executing conflict resolution on E_2' gotten from the inheritance binding and E_1' gotten from the evidence binding.

8.5   Conclusions and Discussions

FLO-based knowledge system framework features decentralized knowledge manipulation and multi-threaded approximate reasoning so that has the advantages of maximizing parallelism in problem solving and maximizing flexibility on mixing knowledge representation schemes and reasoning strategies. Software reuse is also considered during the development of this framework, so the software productivity could be guaranteed. Based on the FLO-based framework, a software development tool, which includes domain knowledge editor, knowledge-level simulator, object model editor, object simulator, and C++ programming editor, has been developed. A user can develop its target application from high-level knowledge level, object model level, or programming language level. The translations among the presentations in various levels are automatic, and simulators do not work on physical code but on logical models so that can support rapid prototyping facility. A class library supporting the automatic code generation of FLO based systems is also implemented in the development tool.

References

1. Bogdan C. et al. Integrating sets, rules, and data in an object-oriented environment. *IEEE Expert*, pages 59–66, 1993.
2. J. Rumbaugh et al. *Object-Oriented Modeling and Design*. Prentice Hall, New Jersey, 1991.
3. David W. Franke. Imbedding rule inferencing in applications. *IEEE Expert*, pages 8–14, 1990.
4. S. Greco, N. Leone, and P. Rullo. COMPLEX: An object-oriented logic programming system. *IEEE Trans. on Knowledge and Data Engineering*, 4:344–359, 1992.
5. Hermann Kaindl. Object-oriented approaches in software engineering and artificial intelligence. *Object-Oriented Programming*, pages 38–45, 1994.
6. Yau-Hwang Kuo and Shiuh-Chu Lee. A tool for building object-oriented blackboard systems. *SoftPro Conference*, C8:1–24, 1992.
7. Ajit Narayanan and Yuanping Jin. Object-oriented representations, causal reasoning and expert systems. *Expert Systems*, 8:13–17, 1991.
8. Stephen T. C. Wong and John L. Wilson. A set of design guidelines for object-oriented deductive systems. *IEEE Trans. on Knowledge and Data Engineering*, 5:895–900, 1993.
9. L. A. Zadeh. Fuzzy sets. *Information and Control*, 8:338–353, 1965.
10. H. J. Zimmermann. *Fuzzy Set Theory and Its Applications*. Kluwer Academic Publishers, Boston, second edition, 1991.

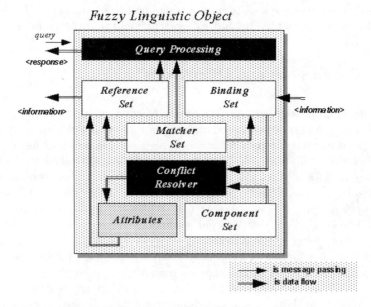

Fig. 8.1. Structure of fuzzy linguistic object.

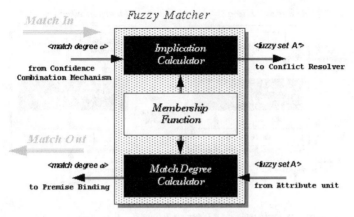

Fig. 8.2. Structure of Fuzzy Matcher.

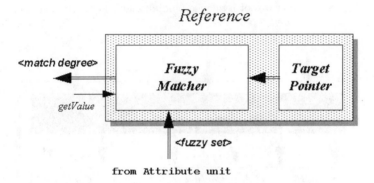

Fig. 8.3. Structure of a Reference.

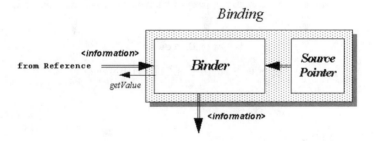

Fig. 8.4. Structure of basic binding element.

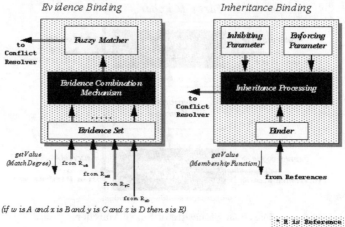

Evidence Binding   *Inheritance Binding*

(if w is A and x is B and y is C and z is D then s is E)

* R is Reference

Fig. 8.5. Structures of Evidence binding and Inheritance binding.

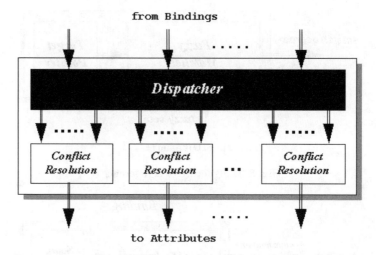

from Bindings

Dispatcher

Conflict Resolution   *Conflict Resolution*   ...   *Conflict Resolution*

to Attributes

Fig. 8.6. Structure of Conflict Resolver.

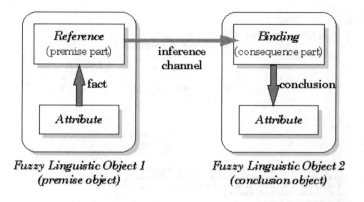

Fig. 8.7. Inference channel between fuzzy linguistic objects

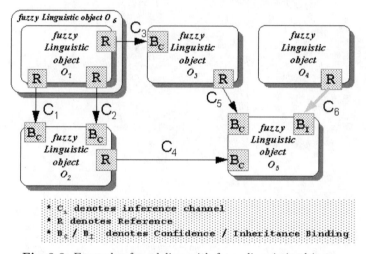

Fig. 8.8. Example of modeling with fuzzy linguistic objects.

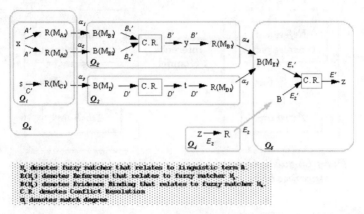

M_A denotes fuzzy matcher that relates to linguistic term A
$R(M_A)$ denotes Reference that relates to fuzzy matcher M_A
$B(M_A)$ denotes Evidence Binding that relates to fuzzy matcher M_A
C.R. denotes Conflict Resolution
α denotes match degree

Fig. 8.9. Inference process of modeling with fuzzy linguistic objects.

9 Fuzzy Evaluation of Domain Knowledge

Bedir Tekinerdoğan and Mehmet Aksit

TRESE Software Engineering group, Dept. of Computer Science, University of Twente, P.O. Box 217, 7500 AE, Enschede, The Netherlands bedir— aksit— @cs.utwente.nl

9.1 Introduction

Software engineering can be essentially seen as a problem solving process that aims to find software solutions for a given problem. The problem is typically initiated by the client's requirement specification and the solution is defined as a software program. Providing a solution for a given problem is not trivial and involves the accumulation and utilization of a huge amount of knowledge. This process of identifying the solution domain knowledge and extracting this knowledge to produce solutions is defined as solution domain analysis [4].

One of the core activities of the solution domain analysis process is the identification of the knowledge sources from which the necessary solution domain concepts will be extracted. To provide quality software it is necessary to elicit the important knowledge sources for a given problem, so that suitable solution abstractions can be identified. As a matter of fact, the quality of the adopted knowledge sources intrinsically defines the quality of the software solution. The corresponding domain knowledge space, though, may be very large and evaluating the knowledge sources may as such be complicated.

This quality of knowledge sources is, on the one hand, determined by its objectivity value, and on the other hand, by its relevance value for the given problem. In practice, evaluation of knowledge domains is uncertain, vague and very often based on the subjective interpretation of the domain engineer. Moreover, the relevance and objectivity values may change due to newly generated knowledge or evolving requirements. Two-valued logic based decisions that either result in accepting or rejecting the knowledge source do not conform with the conceptual evaluation of the human engineer and may easily result in information loss and inappropriate evaluation of the knowledge sources.

We propose to apply fuzzy logic techniques in which knowledge sources are assigned fuzzy linguistic quality values to express the quality degrees. This provides a more precise evaluation and can better cope with the evolution of the knowledge sources and the corresponding problems that are expressed as software requirements. For this purpose we have developed a fuzzy controller to evaluate the relevance and objectivity quality factors of the knowledge sources and determine the abstraction qualities. The fuzzy controller is validated with an experimental case study in which a set of knowledge sources need to be evaluated by a transaction domain expert and a group of novice transaction designers, for two distinct problems.

The remainder of the paper is organized as follows: In section 9.2, we define the background knowledge on solution domain analysis. Section 9.3 defines the problem statement in evaluating knowledge domains. Section 9.4 describes the fuzzy control approach that we have adopted to evaluate knowledge domains. Section 9.5 provides a case study for the design of atomic transaction systems. Finally, section 9.6 provides the related work and finally section 9.7 that provides the conclusions.

9.2   Solution Domain Analysis

Solution domain analysis aims to identify the right solution domains for the given problems and extract the relevant knowledge from these domains to come up with a feasible solution [4,10]. Fig. 9.1 represents a conceptual model for illustrating the solution domain analysis process that we apply.

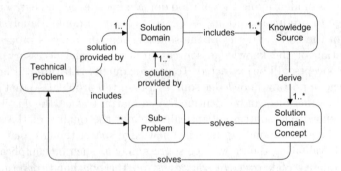

Fig. 9.1. The basic concepts in solution domain analysis

Hereby, the rounded rectangles represent the concepts and the directed arrows represent the associations between these concepts. The figure typically illustrates the relations between the given problem, the solution domain analysis process and the extracted solution domain concepts that forms the output of the solution domain analysis process. The concept *Technical Problem* represents the problem that needs to be solved and likewise forms an input to the solution domain analysis process. For every *Technical Problem* a solution is provided by one or more *Solution Domains*. The concept *Technical Problem* includes zero or more *Sub-Problems*. The concept *Solution Domain* represents the set of *Knowledge Sources* that provide the concepts for solving the problem. From every *Knowledge Source one or more Solution Domain Concepts* can be derived.

For the overall problem and each sub-problem we search for the solution domains that provide the solution abstractions to solve the technical problem.

The solution domains for the overall problem are more general than the solution domains for the sub-problems. In addition, each sub-problem may be recursively structured into sub-problems requiring more concrete solution domains on their turn.

Each identified solution domain may cover a wide range of solution domain knowledge sources. These knowledge sources may not all be suitable and vary in quality. For distinguishing and validating the solution domain knowledge sources we basically consider the quality factors of *objectivity* and *relevance*. The objectivity quality factor refers to the solution domain knowledge sources itself, and defines the general acceptance of the knowledge source. Solution domain knowledge that is based on a consensus of a community of experts has a higher objectivity degree than solution domain knowledge that is just under development. The relevance factor refers to the relevance of the solution domain knowledge for solving the identified technical problem.

The relevance of the solution domain knowledge is different from the objectivity quality. A solution domain knowledge entity may have a high degree of objective quality because it is very precisely defined and supported by a community of experts, though, it may not be relevant for solving the identified problem because it addresses different concerns. To be suitable for solving a problem it is required that the solution domain knowledge is both objective and relevant.

The evaluation of a knowledge source based on the quality factors of relevance and objectivity will result in the quality factor that we term *abstraction quality* [2]. The abstraction quality defines the importance of the corresponding knowledge source for extracting solution concepts. The highest abstraction quality is achieved when both the relevance and the objectivity of the knowledge source are high. A high abstraction quality of the knowledge source means that it provides the fundamental concepts for producing a solution with high quality, that is, a solution that fully meets the requirements and which is stable. The relation between the three quality factors may be given in the following empirical formula:

Abstraction Quality (ks) = (Objectivity(ks), (Relevance(ks))

Hereby *Abstraction Quality(), Objectivity() and Relevance()* represent functions that define the corresponding quality factors of the argument ks, that stands for solution domain knowledge source. For solving the problem, first the solution domain knowledge with the higher abstraction qualities is utilized. The measure of the objec-tivity degree can be determined from general knowledge and experiences. The measure for the relevance factor can be determined by considering whether the identified solution domain source matches the goal of the problem. Note, however, that this formula should not be interpreted too strictly and rather be considered as an intui-tive and practical aid for prioritizing the identified solution domain knowledge sources.

9.3   Problem Statement

A simple approach to evaluate the available knowledge sources for a given problem is to provide Boolean variables *objective, relevant,* and *abstraction quality* for each knowledge source and as such assign either the values true or false to it. Typically we could express the corresponding heuristic rule as follows:

> **IF** knowledge source is *RELEVANT* for the problem **AND** *OBJECTIVE*
>
> **THEN** knowledge source has *ABSTRACTION QUALITY*

If we apply traditional two-valued logic for this rule, then a knowledge source either completely possesses the qualities of relevancy, objectivity and abstraction quality, or it does not. This implies that a knowledge source possesses the abstraction quality only in case it is both considered relevant and objective.

In practice, the process of domain analysis, though, is complex and often related to subjective evaluations, vagueness and uncertainty. Therefore, for a more practical and precise evaluation of the knowledge sources we state that the following three requirements are necessary:

1. *Expressing the degree of quality*
   The evaluation of the objectivity and the relevance value of knowledge domains are basically dependent on the background and expertise of the domain engineer. For a given knowledge source it may be hard to decide whether it completely possesses the quality factors or not. Rather, the domain engineer may decide that it partially possesses the objectivity, relevance and the abstraction quality. Formally, this means that a knowledge source ks of a solution domain SD is mapped to a number in [0,1]. This holds for all the three quality factors:

$$Relevance(ks) : SD \rightarrow [0,1]$$
$$Objectivity(ks) : SD \rightarrow [0,1]$$
$$AbstractionQuality(ks) : (Objectivity(ks), (Relevance(ks)) \rightarrow [0,1]$$

2. *Need for linguistic evaluation of knowledge sources*
   knowledge sources may be evaluated by assigning numbers to their corresponding quality factors. In practice, however, this is counter to the intuition of the domain engineer, which is rather based on linguistic evaluations, such as fairly, substantially, possibly etc. To cope with this, the domain knowledge evaluation approach must therefore provide means to express quality factors using natural linguistic terms to facilitate the communication about their decisions.

3. *Providing means to cope with evolution of knowledge and problems*
   As a matter of fact, knowledge domains are not static but evolve over time. On the one hand, knowledge domains may become obsolete and less useful for solving a problem. On the other hand, they may become

more important, for example, after one has better understood the problem. In addition to the evolution of knowledge, the requirements may evolve as well, and as such the corresponding problem that needs to be solved may change in parallel. Both cases, that is, evolution of knowledge and evolution of problems, may impact the value of the quality factors of the knowledge sources. This implies that the evaluation of the quality of knowledge sources should not be absolute but adaptable to the changing context. Adopting two-valued logic inherently leads to the absolute elimination or acceptance of the knowledge sources and as such fails to cope with this evolution of knowledge and problems appropriately. To address this evolution properly, it is required that knowledge sources are preserved and their quality is adapted to the changing context.

9.4   Fuzzy knowledge Source Evaluator

We believe that the evaluation of knowledge domains may be more effectively supported by the use of fuzzy logic and in particular fuzzy control techniques. For this purpose we have designed a fuzzy control system for evaluating domain knowledge, which is illustrated in Fig. 9.2.

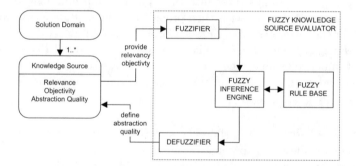

Fig. 9.2. Fuzzy Controller for defining Abstraction Quality of Knowledge Sources

Note that Fig. 9.2 is an elaboration on the model of Fig. 9.1 in that it provides a fuzzy controller, called *Fuzzy Knowledge Source Evaluator (FKSE)*. The FKSE follows the general structure of fuzzy controllers [15] and consists of the four modules *Fuzzifier, Fuzzy Inference Engine, Fuzzy Rule Base, and Defuzzifier*. The basic inputs for FKSE are the values for the *relevance* and the *objectivity* quality factors of the knowledge source, which are used to compute the value for *abstraction quality* of the knowledge source.

The membership functions for the input linguistic variables, *relevance and objectivity*, as well as the output linguistic variable *abstraction quality* are given in Fig. 9.3. From an experimental perspective we have applied the triangular membership functions, though, other membership functions may

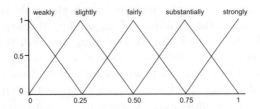

Fig. 9.3. Shape of membership functions of the linguistic input variables Relevance, Objectivity and the linguistic output variable Abstraction Quality

be adopted as well. The optimal membership functions may be determined, for example, after a set of comparative experiments. We will not elaborate on this topic in this paper. For the triangular membership functions of the quality factors we have adopted five linguistic values: *weakly, slightly, fairly, substantially, and strongly.*

The evaluation of the relevance and objectivity quality factors are provided to the module *Fuzzifier*. The evaluation may be both expressed numerically, for example as a crisp value in [0..1] or linguistically using one of the five linguistic values. In the first case, the module *Fuzzifier* takes these crisp values as input and maps these into their membership functions and truth-values. The resulted fuzzy set is then provided to the module *Fuzzy Inference Engine*. In the latter case, the *Fuzzifier* provides the fuzzy set directly to the module *Fuzzy Inference Engine*. The module *Fuzzy Inference Engine* uses the fuzzified values to evaluate the control rules that are stored in the *Fuzzy Rule Base*. The adopted meta-rule in this fuzzy rule base is as follows:

IF knowledge source is <relevance value> *RELEVANT* **and** <objectivity value> *OBJEC-TIVE*

THEN knowledge source has <abstraction quality value> *ABSTRACTION QUALITY.*

This meta-rule can be parameterized with one of the five linguistic values for the quality factors, to define sub-rules. Since, both the *relevance* and *objectivity* quality factors have five linguistic values, we can derive 25 rules from this meta-rule. These rules are represented in Table 9.1.

The first cell of the table, for example, represents the following fuzzy sub-rule:

IF ks is <weakly> *RELEVANT* **and** <weakly> *OBJECTIVE*

THEN ks has <abstraction quality value> *ABSTRACTION QUALITY.*

In Table 9.1, the relevance quality and objectivity quality have equal weight and as such the table is symmetric along the upper left to right bottom diagonal axis. Alternatively, one may consider the *relevance* quality factor more important than the *objectivity* quality factor. In that case, the *relevance* quality factor must have a larger weight than the *objectivity* quality factor. Based on this assumption we developed the fuzzy rules as given in Table 9.2. Note that this table is not symmetric anymore.

Table 9.1. Rules for inferring abstraction quality of knowledge sources with equal weighting factors for the input variables relevance and objectivity

Abstr. Quality	Objectivity				
	WE	SL	FA	SU	ST
WE	WE	WE	SL	SL	FA
SL	WE	SL	SL	FA	FA
FA	SL	SL	FA	FA	SU
SU	SL	FA	FA	SU	SU
ST	FA	FA	SU	SU	ST

(Relevance labels the rows)

The module *Fuzzy Inference Engine* executes the fuzzy rules given the input fuzzy set. In our FKSE we have defined the max-min inferencing method although, other inferencing methods may be adopted equally. We do not further discuss this with respect to the scope of the paper.

Table 9.2. Rules for inferring abstraction quality of knowledge sources with higher weighting factor for relevance factor

Abstr. Quality	Objectivity				
	WE	SL	FA	SU	ST
WE	WE	WE	WE	WE	WE
SL	WE	SL	SL	FA	FA
FA	WE	FA	FA	FA	SU
SU	SL	FA	FA	SU	SU
ST	FA	FA	SU	SU	ST

(Relevance labels the rows)

After the module *Fuzzy Inference* has executed the fuzzy rules it generates a fuzzy output set, which is then provided to the module *Defuzzifier*. The module *Defuzzifier* converts the provided fuzzy set into a crisp value for the *abstraction quality*. For the defuzzification the centroid method is applied.

Once the knowledge sources have been assigned values for *abstraction quality*, the domain engineer can do an explicit trade-off analysis to decide whether the corresponding knowledge source fulfills the required quality to extract the solution domain concepts. An appropriate evaluation will as such improve the quality of the solution abstractions that are derived from selected knowledge sources.

9.5   Case Study: Evaluating Transaction Domain Knowledge

In this section, we will illustrate the application of the fuzzy controller using an experimental case study on the evaluation of the knowledge domains for the design of atomic transaction systems. The goal of the case study is to validate the applicability of our knowledge evaluation approach to real design cases, both for domain experts, and designers who are inexperienced in the corresponding domain.

Section 9.5.1 presents the set of selected knowledge sources that need to be evaluated against two distinct transaction design problems. Section 9.5.2 illustrates the application of the fuzzy controller by adopting the evaluation of the given knowledge sources by a transaction domain expert. Section 9.5.3 adopts and discusses the evaluation of the knowledge sources by novice transaction system designers. Finally, in section 9.5.4 we provide the conclusions of this experimental case study with respect to the earlier defined requirements in section 9.3.

9.5.1   Selection of Problem and Knowledge Sources

We can derive a large number of publications on the theory of transaction systems. Informally atomic transactions are characterized by two properties: serializability and recoverability [8]. Serializability means that the concurrent execution of a group of transactions is equivalent to some serial execution of the same set of transactions. Recoverability means that each execution appears to be all or nothing; either it executes successfully to completion or it has no effect on data shared with other transactions.

Many different transaction systems can be designed by extracting various concepts from the transaction literature. Although a common architecture for transaction systems can be derived, transaction systems may nevertheless differ in the selected transaction protocols, such as transaction management, concurrency control protocols, recovery protocols, and data management techniques. Obviously, every transaction system design may need its own dedicated knowledge. The kind of transaction system is basically defined by the corresponding problems. In our experimental case study we have adopted the following two distinct design problems.

- *Example Problem 1:*
  Design a transaction system for a distributed system which is based on a two-phase locking concurrency control scheme and a recovery scheme that handles transaction failures based on image logging. The transaction management needs to consider only flat transactions.

- *Example Problem 2:*
  Design an advanced transaction system with adaptable transaction properties. In addition to flat transactions, the system must be able to compose

*nested transactions as well. The choice of the concurrency control and the
recovery protocols must be made adaptable according to the performance
characteristics of the system.*

Note that the first design problem requires the design of a transaction
system with fixed properties while the second design problem requires adaptable transaction protocols. To provide a solution for the first problem requires
knowledge on the specific protocols, whereas the second problem requires
knowledge on a wide range of transaction protocols and additionally it requires knowledge on performance of transaction protocols, and adaptation
protocols to switch between the various protocols.

The knowledge sources that needed to be evaluated are presented in Table
9.3. We have deliberately selected knowledge sources that differ from each
other to capture the impact of the fuzzy reasoning process. Differences can
be observed with respect to the form of the knowledge source, the date of
publication, the location of the publication, the generality of the publication
etc.

Table 9.3. A selected set of knowledge sources for the overall solution domain

KS	KNOWLEDGE SOURCE
KS1	Concurrency Control & Recovery in Database Systems[7]
KS2	Atomic Transactions[17]
KS3	An Introduction to Database Systems[11]
KS4	Database Transaction Models for Advanced Applications[12]
KS5	The design and implementation of a distributed transaction system based on atomic data types[25]
KS6	Concurrency Control Performance Modeling: Alternatives and Implications[1]
KS7	Principles of Transaction Processing[8]
KS8	Course Notes of Transaction Design
KS9	Concurrency Control in Advanced Database Applications[6]
KS10	Conference Proceedings on Advanced Transaction Systems and Applications
KS11	On-Line Transactions Tutorial[21]
KS12	Transaction Domain Expert with 15 years of experience
KS13	Design of Adaptable Transaction Systems[22]
KS14	Nested Transactions[18]
KS15	Adaptable Concurrency Control for Atomic Data Types[5]
KS16	A survey of techniques for Synchronization and Recovery in Decentralized Computer Systems[16]

9.5.2   Evaluation by a Domain Expert

Table   9.4 represents the evaluation of *relevance* and the *objectivity* of the
knowledge sources by a transaction domain expert who has experience in the
theory, design and implementation of a wide range of transaction systems.

During the evaluation all the knowledge sources were actually made available and could be analyzed. The knowledge sources have been evaluated for

Table 9.4. The evaluation of the relevance and objectivity of knowledge sources by a transaction domain expert

KS	RELEVANCE P1	RELEVANCE P2	OBJECTIVITY
1.	ST	SU	SU
2.	FA	FA	ST
3.	SL	WE	FA
4.	WE	ST	SU
5.	SU	SL	ST
6.	WE	ST	ST
7.	SL	WE	FA
8.	SL	WE	SL
9.	FA	ST	SU
10.	WE	FA	SU
11.	WE	WE	FA
12.	ST	ST	ST
13.	FA	SU	FA
14.	FA	SU	ST
15.	WE	ST	SU
16.	SU	WE	WE

both problem 1 and problem 2. Since *relevance* is dependent on the problem, the table provides one row for the relevance of each of the two problems. In contrast, *objectivity* is problem independent and as such includes only one row.

Using the input values for the relevance and objectivity quality factors, the fuzzy knowledge evaluator can infer the abstraction quality for each knowledge source. To illustrate the inference mechanism we consider, for example, the inference of the abstraction quality for knowledge source KS4 for problem 1. For this knowledge source the fuzzy relevance value is *weakly*, for problem 1, and its objectivity value is *substantially*. Considering the membership functions as illustrated in Fig. 9.3, we can derive that for the linguistic variable *relevance*, the fuzzy value *weakly* overlaps with fuzzy value *slightly*. For the linguistic variable *objectivity*, the input value *substantially* overlaps with the values *fairly* and *strongly*. The inference engine will fire all rules but in the end only those rules for which *relevance* is *weakly* or *slightly*, and *objectivity* is *fairly, substantially* or *strongly*, will have an impact on the final result. For the experimental case study we adopted the rules as defined in Table 9.1, that is, *relevance* and *objectivity* of knowledge sources have equal weight. As such, for determining the fuzzy set of the abstraction quality of KS4, the following six rules will have a impact:

1. **IF** ks is <weakly> *RELEVANT* **and** <fairly> *OBJECTIVE*
   THEN ks has <slightly> *ABSTRACTION QUALITY*.
2. **IF** ks is <weakly> *RELEVANT* **and** <substantially> *OBJECTIVE*
   THEN ks has <slightly> *ABSTRACTION QUALITY*.

3. **IF** ks is <weakly> *RELEVANT* **and** <strongly> *OBJECTIVE*
   THEN ks has <substantially> *ABSTRACTION QUALITY.*
4. **IF** ks is <slightly> *RELEVANT* **and** <fairly> *OBJECTIVE*
   THEN ks has <slightly> *ABSTRACTION QUALITY.*
5. **IF** ks is <slightly> *RELEVANT* **and** <substantially> *OBJECTIVE*
   THEN ks has <fairly> *ABSTRACTION QUALITY.*
6. **IF** ks is <slightly> *RELEVANT* **and** <strongly> *OBJECTIVE*
   THEN ks has <fairly> *ABSTRACTION QUALITY.*

The execution of these rules results in the fuzzy set that is illustrated in
Fig. 9.4a.

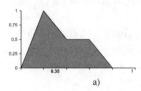   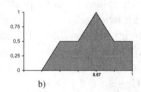

a)   b)

Fig. 9.4. Fuzzy set of the inferred abstraction qualities for knowledge source 4 for
a) problem 1 and b) problem 2

The execution of the fuzzy rules for problem 2 will yield the fuzzy set as
given in Fig. 9.4b. Note that for problem 2 the set of rules that provide impact
on the final result will be different than for problem 1 because of the different value for *relevance*. Using the centroid method the module *Defuzzifier*
computes a crisp value of the fuzzy sets, which are 0.35 and 0.67 for problem
1 and problem 2, respectively. These numbers give the software engineer a
practical indication of the quality of the knowledge source. In this case, KS4
has a clearly higher abstraction quality for problem 1 than for problem 2.

Similar to KS4, the fuzzy knowledge evaluator yields the fuzzy sets for
the other knowledge sources. Fig. 9.5. shows the defuzzified values of the
abstraction qualities of the sixteen knowledge sources, both for problem 1
and problem 2. This figure gives already an hint of the quality of the various
knowledge sources and this information is valuable for the software engineer
who needs to extract the abstractions from the knowledge sources to develop
the solution. The lines in the figure do not have a specific meaning but have
been only included for visibility. Let us now take a closer look at Fig. 9.5 and
interpret the resulted abstraction qualities.

A global look at the figure shows, that knowledge sources that only deal
with advanced transactions (4, 9, 10), performance modeling (6) and dynamic
adaptation (13,15) have got a low abstraction quality for problem 1, that is,
the design of a flat transaction system with fixed properties.

Knowledge source 3 has got a low abstraction quality for both problems.
This may be because it is not directly or explicitly related to the subject of
transaction systems.

Knowledge source 5 has the highest abstraction quality for problem 1 but
a lower value for problem 2. This may be attributed to the fact that it is

both a journal paper, leading to a strong objectivity value, and because it is directly related to the design of flat transaction systems.

Knowledge source 6, the journal paper on performance modeling, has the highest abstraction quality for problem 2 but a rather low abstraction quality for problem 1. This may be due to its irrelevance to problem 1. It is not the lowest abstraction quality for problem 1 because it has been evaluated as strongly objective, which indirectly increases the abstraction quality.

Knowledge source 12, the transaction domain expert, has the highest abstraction quality for both problems. This may be explained from the fact that the expert knows both problems well and possesses recent and strongly objective knowledge.

Knowledge source 13, a MSc thesis on the design of adaptable transaction systems, although strongly relevant for problem 2, has not got the highest abstraction value. This may be because it is not recent and the fact that it is a MSc thesis.

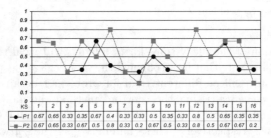

KS	1	2	3	4	5	6	7	8	9	10	11	12	13	14	15	16
P1	0.67	0.65	0.33	0.35	0.67	0.4	0.33	0.33	0.5	0.35	0.33	0.8	0.5	0.65	0.35	0.35
P2	0.67	0.65	0.33	0.67	0.5	0.8	0.33	0.2	0.67	0.5	0.33	0.8	0.5	0.67	0.67	0.2

Fig. 9.5. Inferred (defuzzified) abstraction qualities of the knowledge sources after the evaluation by a domain expert

The above observations show that the evaluation of the knowledge sources using fuzzy linguistic rules is not ad hoc and reasonably match the intuition. In addition, we can observe that the abstraction qualities have indeed been evaluated differently for both problems, and this difference also provides a sound conceptual interpretation. The knowledge sources have now been more precisely evaluated than in case of a two-valued logic evaluation approach, because every evaluation closely matches the intuition and as such no information is lost.

9.5.3   Evaluation by Novice Transaction System Designers

To determine the validity and the applicability of the heuristic fuzzy rules for inexperienced domain engineers, we have presented the 16 knowledge sources in Table 9.3, also to novice transaction system designers. This group consisted of 25 fourth year students of Computer Science at the University of Twente, in the lecture on object-oriented software analysis and design. The students got, a week before, a one and half-hour lecture on solution domain analysis and evaluation of knowledge sources based on the relevance and objectivity quality factors. Similar to the evaluation by the transaction domain expert, the students got the actual knowledge sources during the experimental case study, and evaluated these one by one, separately and independently. The students did not know in advance, though, that they were involved in an experimental case study, but considered the evaluation process more as a practical assignment. The evaluation of the knowledge source took about two hours and consisted again of assigning the five fuzzy values of *weakly, slightly, fairly, substantially,* and *strongly* to the corresponding knowledge sources. We have collected and processed all the results of all students and did not eliminate any.

For every knowledge source we have counted the frequency of the five linguistic fuzzy values and represented these in histograms as illustrated in the Appendix. For example, for the relevance of knowledge source 1 for problem 1, none student assigned the values *weakly* and *slightly*, 2 students assigned the value *fairly*, 12 students thought that it was *substantially*, and 11 students assigned the value *strongly*.

We have implemented an algorithm for deriving the fuzzy sets of a given histogram. Hereby, the average of the 25 fuzzy values in each histogram is taken to yield a new fuzzy set. Consider for example the computation of the average fuzzy set for the relevance of knowledge source 1 (KS1) for problem 1. The histogram of KS1 can be fuzzified by taking the average of 2 *fairly*, 12 *substantially*, and 11 *strongly* fuzzy values. The resulted fuzzy set is shown in Fig. 9.6. Fuzzy set for the relevance of knowledge source 1 for problem 1, derived from the histogram of the evaluation of novice transaction system designers

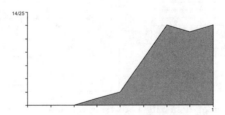

Fig. 9.6. Fuzzy set for the relevance of knowledge source 1 for problem 1, derived from the histogram of the evaluation of novice transaction system designers

We have done this for each histogram and derived the fuzzy sets of the relevance and objectivity quality factors of each knowledge source. The fuzzy sets have then been provided as an input to the fuzzy knowledge evaluator to provide the abstraction qualities. The results of the defuzzified sets of the abstraction qualities are shown in Table 9.5.

Table 9.5. Inferred (Defuzzified) Abstraction Quality of a group of novice transaction designers

KS	DEFUZZIFIED ABSTR. QUAL P1	DEFUZZIFIED ABSTR. QUAL P2
1.	0.67	0.52
2.	0.65	0.62
3.	0.33	0.42
4.	0.35	0.50
5.	0.67	0.50
6.	0.40	0.53
7.	0.33	0.58
8.	0.33	0.44
9.	0.50	0.51
10.	0.35	0.62
11.	0.33	0.49
12.	0.80	0.58
13.	0.50	0.55
14.	0.65	0.53
15.	0.35	0.55
16.	0.35	0.43

To interpret and validate these results we will provide a comparison with the evaluation of the transaction domain expert. Fig. 9.7 shows a comparison between the evaluation of the inferred evaluation values for the domain expert and the novice group.

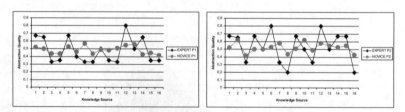

Fig. 9.7. Comparison of defuzzified abstraction qualities of domain expert and novice transaction designers for (a) problem 1 and (b) problem 2

A first glimpse at Fig. 9.7 makes clear that the novice group is more careful in their evaluation than the transaction domain expert, because for both problems the difference among the abstraction qualities of the various knowledge sources is less than in the evaluation by the domain expert. This may be attributed to the self-confidence of the domain expert who has a complete and objective overview of the domain and as such can make more sharp decisions. The students on the other hand may be more careful if they are not sure about their decisions.

If we take a look at the histograms we can observe that there is a reasonable consensus among the 25 students for the evaluation of the knowledge sources. For some knowledge sources (such as 1, 6, 12, 14 and 15) this is more obvious, than others (such as 2, 5, 11, 16). The lack of consensus may show the difficulty of the evaluation, which is influenced by the background and experience of the individual students.

The experience factor of the novice group is obvious for the evaluation of knowledge source 2, a textbook including many formal algorithms and proofs on basically nested transactions. For the relevance quality for problem 1 we can observe that the students do not form a common opinion. This may be attributed due to their lack of experience in interpreting and applying formal algorithms. The values for the objectivity, however, are commonly assigned high values. The reason for this may be that although they cannot precisely evaluate the relevance they have confidence in its scientific objectivity because of the many formulas.

The students were able to distinguish the relevancy of the knowledge sources for both problems. If we compare the defuzzified abstraction qualities with that of the abstraction qualities inferred from the domain expert's evaluation we can observe that the difference in the quality values are reasonably similar. For example, for KS6 the evaluation of the domain expert resulted in the values of 0.4 and 0.8 (Fig. 9.5), for problem 1 and problem 2, respectively. The evaluation by the novice group resulted in the same increasing order for the abstraction quality of 0.4 and 0.53 (Table 9.5), for problem 1 and problem 2, respectively. Although, the domain expert's evaluations are more sharp, the application of the fuzzy heuristic rules resulted generally in the same ordering of the abstraction qualities of the knowledge sources. As a matter of fact, this ordering of the knowledge sources is of higher importance than the assigned values.

We may infer many more conclusions from this experimental case study, though, due to space limitations we will not elaborate on it.

9.5.4 Conclusions of the Overall Case Study

The goal of the case study was to validate the applicability of our knowledge evaluation approach with respect to the identified requirements. In the previous two sections we have respectively discussed the evaluations of the domain expert and the novice group individually. For both cases we have shown that the fuzzy knowledge evaluator is useful and applicable.

First of all, the use of linguistic values for the assessment of the knowledge sources was applied quite straightforward; the evaluation of knowledge sources was done in a rather short period (about 2 hours) and every student could evaluate the knowledge sources independently.

Instead of a classification of accepted and rejected knowledge source as it would result from using two-valued logic, there is a clear graded ordering of the knowledge sources, which likewise provide a more precise evaluation. This is valuable for selecting the knowledge sources for extracting the solution abstractions.

The evaluation of both the domain expert and the novice group illustrates the shift in the values for the abstraction qualities for problem 1 and problem 2. Our approach does not need to eliminate knowledge sources and as such the abstraction qualities may be changed according to the evolution of the context, that is, problems and knowledge.

The reasonable ordering of the inferred abstraction qualities from the domain expert and the novice group show that the fuzzy rules may be successfully applied. The novice group was more careful and had more problems in the evaluation, possibly due to the lack of general scientific experience.

9.6   Related Work

The solution domain analysis process basically forms the core of our earlier work on the *synthesis-based design process* [23]. Synthesis is a problem solving approach that is applied in mature engineering disciplines such as electrical engineering, mechanical engineering and chemical engineering. The synthesis-based design process consists basically of the sub-processes of technical problem analysis, solution domain analysis and alternative design space analysis. The technical problem analysis phase aims to define the technical problems that have been initiated by the requirement specification. The solution domain analysis aims to find the corresponding solution domains for the given problem and extracts solution domain concepts from these domains. The alternative design space analysis aims to define the space of the alternatives for the given problem and evaluates these against quality criteria. This paper focuses on the solution domain analysis process of the synthesis-based design process. We aim to formalize the whole synthesis process and apply fuzzy control were necessary. The next step to integrating fuzzy control in the synthesis-based design process will be the evaluation of the extracted solution abstractions.

Fuzzy control has been applied in many different fields but very few of them in the area of software engineering. In [3] fuzzy control is adopted to enhance object-oriented methods by modeling and controlling the design alternatives. The authors maintain that design alternatives during the software development process needs to be preserved to allow further refinements. Two-valued logic is not able to meet this requirement and eliminates al-ternatives too early.

Several domain analysis processes have been published, e.g. [14,19,20,10]. Two surveys of various domain analysis can be found in [4,24]. In [10], a more recent and extensive up-to-date overview of domain engineering methods is provided. The fuzzy evaluation approach in this paper can be applied to all of these domain analysis methods.

Solving a problem requires first identifying the corresponding solution domains. This may a difficult task if the domain knowledge space is very large. To support the search for the right solution domains we may categorize the domain knowledge [13]. Software engineering applies knowledge of a wide range of application domains, one of them especially is the Computer Science domain, such as programming languages, operating systems, analysis and design methods etc. This type of knowledge has been recently compiled in the so-called Software Engineering Body of knowledge [9].

9.7 Conclusion

Software engineering is a problem solving process in which domain knowledge needs to be applied to provide software solutions. The quality of the produced software solution is intrinsically related to the selected knowledge sources. It is therefore necessary to elicit the important knowledge sources for a given problem, so that suitable solution abstractions can be identified. To be useful for solving a problem, the knowledge source must be relevant to the problem and have objective quality, which together define the so-called abstraction quality. A simplistic approach to evaluate domain knowledge is by using two-valued logic, whereby a knowledge source inherently either has the abstraction quality or not. In practice, however, the software engineer may decide that it partially possesses the abstraction quality and need to express these in natural language. In addition, the evaluation of the knowledge sources cannot be absolute but need to change along with the evolution of the available knowledge and the given technical problems. Adopting two-valued logic leads either to the absolute elimination or preservation of the identified knowledge sources.

To cope with these requirements, we have provided a model and an approach for evaluating domain knowledge using fuzzy logic techniques, the so-called fuzzy knowledge source evaluator (FKSE). The FKSE takes as input the relevance and the objectivity (fuzzy) values of a knowledge source and computes the abstraction quality. The inference engine adopts 25 fuzzy heuristic rules.

We have illustrated the application of FKSE in an experimental case study on evaluating domain knowledge for the design of atomic transactions. Thereby, we have provided two distinct problems to a domain expert and a group of novice transaction system designers. From the case study we concluded that applying fuzzy logic techniques in evaluating domain knowledge is of practical use and can support the solution domain analysis process.

9.8   Acknowledgements

We would like to thank Pim van den Broek for providing us his Java implementation of the fuzzy algorithms, and Lodewijk Bergmans for helping to extend the Java implementations for deriving fuzzy sets from frequency histograms. Further, we thank the students who have cooperated in the experimental case study on evaluating knowledge sources for the transaction domain. This research is supported by the Dutch Scientific Organization (NWO).

References

1. R. Agrawal, M. Carey, and M. Livney. Concurrency control performance modelling: Alternatives and implications. *ACM Transactions on Database Systems*, 12(4):609–654, December 1987.
2. M. Aksit. *Course Notes: Designing Software Architectures*. Post-Academic Organization, 2000.
3. M. Aksit and F. Marcelloni. *Deferring Elimination of Design Alternatives in Object-Oriented Methods*. Concurrency Practice and Experience, 2000.
4. G. Arrango. Domain analysis methods. In R. Schafer, R. Prieto-Diaz, and M. Matsumoto, editors, *Software Reusability*. Ellis Horwood, New York, 1994.
5. M.S. Atkins and M.Y. Coady. Adaptable concurrency control for atomic data types. *ACM Transactions on Computer Systems*, 10(3):190–225, August 1992.
6. N.S. Barghouti and G.E. Kaiser. Concurrency control in advanced database applications. *ACM Computing Surveys*, 23(3), September 1991.
7. P.A. Bernstein, V. Hadzilacos, and N. Goodman. *Concurrency Control & Recovery in Database Systems*. Addison Wesley, 1987.
8. P.A. Bernstein and E. Newcomer. *Principles of Transaction Processing*. Morgan Kaufman Publishers, 1997.
9. P. Bourque, R. Dupuis, A. Abran, J.W. Moore, and L. Tripp. The guide to the software engineering body of knowledge. 16(6):35–45, November/December 1999.
10. Czarnecki and U. Eisenecker. *Generative Programming*. Addison-Wesley, 2000.
11. C.J. Date. *An Introduction to Database Systems*, volume 3. Addison Wesley, 1990.
12. A.K. Elmagarmid (Ed.). *Database Transaction Models for Advanced Applications Transaction Management in Data-base Systems*. Morgan Kaufmann Publishers, 1992.
13. R.L. Glass and I. Vessey. Contemporary application-domain taxonomies. *IEEE Software*, 12(4), July 1995.
14. K. Kang, S. Cohen, J. Hess, W. Nowak, and S. Peterson. Feature-Oriented Domain Analysis (FODA). Technical report, Software Engineering Institute, Carnegie Mellon University, 1990.
15. G.J. Klir and B. Yuan. *Fuzzy Sets and Fuzzy Logic: Theory and Applications*. PrenticeHall, 1995.
16. Kohler. A survey of techniques for synchronization and recovery in decentralized computer systems. *ACM Computing Surveys*, 13(2), 1981.

17. N. Lynch, M. Merrit, W. Weihl, and A. Fekete. *A. Atomic Transactions*. Morgan Kaufmann Publishers, 1994.
18. J.E.B. Moss. *Nested Transactions : an approach to reliable distributed computing*. Cambridge, MA: MIT Press, 1985.
19. R. Prieto-Diaz and G. Arrango (Eds.). *Domain Analysis and Software Systems Modeling*. IEEE Computer Society Press, Los Alamitos, California, 1991.
20. M. Simos, D. Creps, C. Klinger, L. Levine, and D. Allemang. *Organization Domain Modeling (ODM) Guidebook*. 1996.
21. B. Tekinerdoğan. *On-line transactions Tutorial, web-document: www.cs.utwente.nl/ bedir/ TransactionsTutorial*. Dept. of Computer Science, University of Twente.
22. B. Tekinerdoğan. *Design of an Object-Oriented Framework for Atomic Transactions*. MSc. Thesis, Dept. of Computer Science, University of Twente, 1994.
23. B. Tekinerdoğan. *Synthesis-Based Software Architecture Design*. PhD Thesis, Dept. Of Computer Science, University of Twente, March 2000.
24. S. Wartik and R. Prieto-Diaz. Criteria for comparing domain analysis approaches. *In International Journal of Software Engineering and Knowledge Engineering*, 2(3):403–431, 1992.
25. Z. Wu, R.J. Stroud, K. Moody, and J. Bacon. The design and implementation of a distributed transaction system based on atomic data types. *Distributed Syst. Engineering*, 2:50–64, 1995.

10 Application of Fuzzy Rule Extraction to Minimize the Costs of Misclassification on Software Quality Modeling

Zhiwei Xu[1] and Taghi M. Khoshgoftaar[2]

[1] azx095@motorola.com
  Motorola Labs, Motorola Inc.
  Schaumburg, IL 60319,USA
[2] taghi@cse.fau.edu
  Florida Atlantic University
  Boca Raton, FL 33431, USA

10.1 Introduction

Today's modern world has become highly dependent on computers, even on a daily basis. The dependability and functionality of many systems rely on computers, especially its software systems. Consequently, high software quality and reliability is crucial to software developers, software users, and society in general. This is because poor software quality can threaten safety, risk a company's business, or alienate potential customers. It is not advisable to wait until a product's release to consider its software quality, because it may be too late to significantly improve the system.

The field of software metrics is predicated on the idea that measurement of product attributes during development will give insight into the quality of the operational software [5]. Prior research has shown that software product and process metrics collected prior to the test phase can be the basis for fault predictions. A fault is classically defined as a defect that causes a software failure. We intend to enhance the quality of modules recommended by a model early enough to prevent problems due to faults discovered later in the life cycle. This is because it is much more cost-effective to correct software faults early in the development process than later, when they can cause failures [10,14,17]. Therefore, it would be of great benefit to identify those components that are likely to have a high error rate. Management may use these predictions to guide the decision making process for future projects.

Predicting the exact number of faults in each program module is often not necessary, and instead previous research has focused on classification models to identify fault-prone and not fault-prone modules [1,16,18,19]. Such models are based on modeling techniques like nonparametric discriminant analysis. However, before reliability problems become evident, it is difficult to build an explicit quality estimation model at the time of modeling. In such cases, using a *Rule Based Fuzzy Classification* (RBFC), is more appropriate. Christof Ebert [3,4] applied expert knowledge in his fuzzy classification model. However, expert knowledge is often too general to fit a particular data set because different data sets have different characteristics. This stimulates us to seek a way to generate rules from data directly, and show the useful relationship between the independent and dependent variables in a particular data set.

Recently, various methods have been proposed for automatically generating fuzzy if-then rules from numerical data. Most of these methods applied iterative learning procedures or complicated rule generation mechanisms. These algorithms include gradient descent learning methods [7,8,25], genetic algorithm-based methods [11,21], least squares methods [29,30], fuzzy c-means method [33], and fuzzy-neuro method [2,24,26].

In this study, we propose a new way to generate if-then rules. The statistics on the distribution of the independent variables are used to determine membership functions experimentally, and the if-then rules are generated in such a way so as to suppress the Type II misclassification errors [13,15,14]. A Type II misclassification error occurs when a fault-prone module is identified as not fault-prone. We summarize the concepts of the proposed RBFC and its usage in software reliability classification models. To our knowledge, this is the first study that introduces RBFC in software reliability engineering.

The use of the proposed RBFC modeling technique is illustrated using three case studies. Further more, two case studies are used to evaluate the performance of the RBFC models with those built using nonparametric discriminant analysis. Our previous studies used nonparametric discriminant analysis to build classification models for these two case studies, and hence we have compared RBFC models with the nonparametric discriminant analysis models. In the third case study, which is a more recent case study, we also calibrated classification models using another machine learning technique, case-based reasoning [22]. Consequently, we compare the proposed RBFC modeling technique with case-based reasoning.

It was observed that the RBFC models give management more flexible reliability enhancement strategies than the models built using nonparametric discriminant analysis and case-based reasoning. Further more, for the three case studies presented, RBFC yielded more accurate classification results than the corresponding nonparametric discriminant analysis and case based reasoning models.

10.2   Fuzzy Classification

A fuzzy model is a set of if-then rules that maps inputs to outputs. Basically, an RBFC provides an effective way to present the approximate and imprecise nature of the real world. In particular, RBFC appears useful when the systems are not suitable for analysis by conventional quantitative techniques or when the available information on the systems is uncertain or inaccurate.

Most quality factors are directly measurable only after software has become operational, for example, the number of faults. In contrast, most software product and process metrics can be measured during development. The relationship between the software metrics and the quality factors is of course not clear. The goal of RBFC is to generate the rules to express this relationship, and the rules can also be used to guide further design of the systems. In this section, we present an RBFC system used in this study. Basically, RBFC systems are composed of four principal components: a fuzzification facility, a rule base, an inference engine and a defuzzification facility. Fig. 10.1 illustrates the structure of a fuzzy rule-based system.

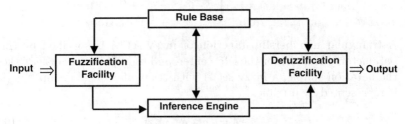

Fig. 10.1. General Structure of a Fuzzy Rule-based System

10.2.1   Fuzzification facility

The fuzzification facility conducts a mapping that converts the crisp values of input variables into a fuzzy singleton. Before the conversion occurs we need to define the membership function for each input and output variable. We consider a single-output fuzzy rule-based system with m-dimensional inputs. Suppose $\{\mathbf{x}_i \mid i = 1, 2, ..., n\}$ and $\{y_i \mid i = 1, 2, ..., n\}$ are the independent and dependent variables, respectively. The training data to generate rules can be represented by n input-output pairs $(\mathbf{x}_i, y_i)$, where $i = 1, 2, ...n$. The independent variable vector $\mathbf{x}_i = (x_{i1}, x_{i2}, ..., x_{im})$ is of size m while the dependent variable y_i is of size one. We define membership functions for each of these m input variables as $\{A_j \mid j = 1, 2, ..., m\}$. And we assume that the domain interval of the jth input variable is divided into K_j fuzzy sets labeled as $A_{j1}, A_{j2}, ..., A_{jK_j}$, for $j = 1, 2, ..., m$. Then the m-dimensional input space is divided into $K_1, K_2, ..., K_m$ fuzzy subspaces:

$$(A_{1p_1}, A_{2p_2}, ..., A_{mp_m}), \quad p_1 = 1, 2, ..., K_1; \quad ...; \quad p_m = 1, 2, ..., K_m \qquad (10.1)$$

There are numerous membership functions, and we list only those we use in our study for fuzzy set A as follows:

- Gaussian membership function:

$$\mu_A(x) = e^{\frac{-(x-c)^2}{2\sigma^2}} \qquad (10.2)$$

- Generalized bell membership function:

$$\mu_A(x) = \frac{1}{1+ \mid \frac{x-c}{a} \mid^{2b}} \tag{10.3}$$

- Triangular membership function:

$$\mu_A(x) = \begin{cases} 0 & x \leq L \\ \frac{x-L}{C-L} & L \leq x \leq C \\ \frac{R-x}{R-C} & C \leq x \leq R \\ 0 & x \geq R \end{cases} \tag{10.4}$$

A triangular membership function of fuzzy set A is specified by three parameters (L, C, R) for the left, center, and right points. In order to keep the notation simple, a fuzzy set A with triangular membership function is represented by a triple.

$$A = (a^L, a^C, a^R) \tag{10.5}$$

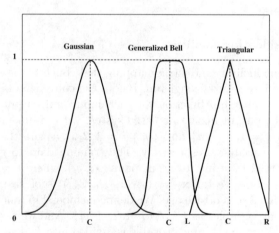

Fig. 10.2. Membership Functions

The plots of different membership functions are shown in Fig. 10.2. In order to simplify the process to determine the membership functions, some fuzzy applications choose evenly-divided membership functions. However, we found that this approach is not able to show the features of a given data. So we use the statistics on the distribution of input variables to define the membership functions. In other words, we do not divide the domain of input variable evenly. Since the variables are likely to cluster around the mean, these points need to be divided into more pieces. We call this method *partial partition*. Fig. 10.3 illustrates the method.

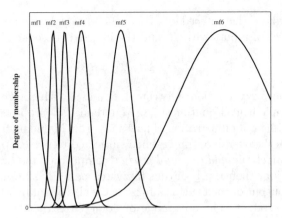

Fig. 10.3. Membership Functions

10.2.2   Rule Base

The rule base contains a set of fuzzy if-then rules. In this study, we use the
if-then rules in the following form:

Rule $R_{p_1...p_m}$: If x_{i1} is A_{1p_1} ... and If x_{im} is A_{mp_m} then y_i is $b_{p_1...p_m}$ (10.6)

where $p_1 = 1, 2, ..., K_1$;   ... ; $p_m = 1, 2, ..., K_m$, and $b_{p_1...p_m} \in \{0, 1\}$ and
$R_{p_1...p_m}$ is the label of the fuzzy if-then rule.

Fuzzy rules can be generated from the fuzzy subspace described in Equation 10.1. Therefore, the number of fuzzy rules should be equal to the number
of the fuzzy subspaces in Equation 10.1, that is the product of $K_1, K_2, ..., K_m$.
However, in our cases studies, some rules do not activate at all, we call these
rules "inactive rules". In order to speed up the reasoning process and reduce
the size of the rule base, we will remove these inactive rules from the rule
base.

10.2.3   Fuzzy inference

The *fuzzy inference system* [9] is a popular computing framework based on
the concepts of fuzzy set theory and fuzzy rules such as $(A \text{ AND } B) \to$
C, where the *antecedent* $A(x)$ and $B(y)$ may be fuzzy sets, the *consequent*
may be a fuzzy set $C(z)$ and the *inference rule* is a relation $R(x, y, z)$ whose
membership function is defined as,

$$\mu_R(x, y, z) = \mu_A(x) \times \mu_B(y) \times \mu_C(z) \qquad (10.7)$$

There are different manifestations of the conjunction operation $\times$. Mamdani
[23] proposed using *minimum* operator for conjunction, and Larsen used *algebraic product* operator for conjunction. Execution of a rule is based on

composition where the antecedent A' is a *fuzzy match* of inputs with A, B' is a *fuzzy match* of inputs with B, and the inferred consequent is

$$C' = (A' \cap B') \circ R \tag{10.8}$$

where R is the fuzzy relation between A, B and C. There are different approachs to implement Equation 10.8. Mamdani [23] proposed a *max-min* method. Fig. 10.4 is an illustration of how a two-rule Mamdani fuzzy inference system derives the overall output z when subjected to two crisp inputs x and y. If we used algebraic *addition and product* composition instead of the max-min composition, then the resulting fuzzy reasoning is shown in Fig. 10.5. The inferred output of each rule is a fuzzy set scaled down by its fuzzy match via algebraic product. The advantage of doing so is that the shape of the original membership function can be preserved, and as a result, the output of the reasoning balances all the information brought by the antecedents and derives the result accordingly [31].

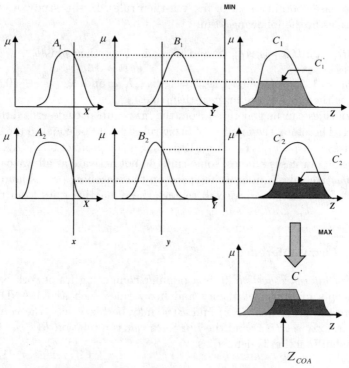

Fig. 10.4. The Mamdani fuzzy inference system using *max-min* method

We can extend the above two inputs system to multiple inputs $\mathbf{x}_i = (x_{i1}, x_{i2}, ..., x_{im})$, the inferred output y_i is defined by

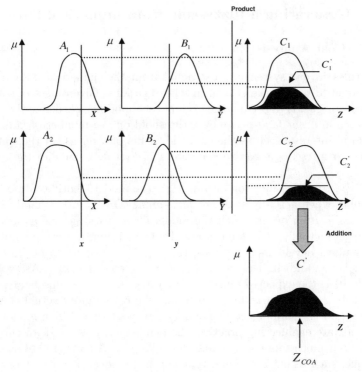

Fig. 10.5. The Mamdani fuzzy inference system using *addition-product* method

$$Y'(\mathbf{x}_i) = \bigvee_{p_1=1}^{K_1} \cdots \bigvee_{p_m=1}^{K_m} [\mu_{p_1\ldots p_m}(\mathbf{x}_i)] \times b_{p_1\ldots p_m} \qquad (10.9)$$

where $\mu_{p_1\ldots p_m}$ is the degree of compatibility of the input vector $\mathbf{x}_i$ to the fuzzy rule $R_{p_1\ldots p_m}$, and it can be obtained by

$$\mu_{p_1\ldots p_m} = \mu_{1p_1}(x_{i1}) \times \cdots \times \mu_{mp_m}(x_{im}) \qquad (10.10)$$

The fuzzy operation $\times$, as discussed before, can be the minimum or algebraic product, and $\bigvee$ can be the maximum or algebraic addition.

10.2.4   Defuzzification Facilities

Defuzzification is a mapping from a space of fuzzy output over an output universe of discourse into a space of crisp output.

10.3   Generating a fuzzy rule from numerical data

10.3.1   Cost Analysis

Fuzzy rules in our system are extracted from the numerical data in such a way so as to reduce the cost of misclassifications. Given a set of software modules, our goal is to predict the class of each module. We define the classes *fault-prone* and *not fault-prone* by a threshold on the number of faults over a period of interest, such as the operational life of a release of the software. This section presents a way to generate a fuzzy rule to predict the module class.

A Type I misclassification occurs when the model identifies a module as *fault-prone* when actually, it is *not fault-prone*. On the other hand, a Type II misclassification occurs when the model identifies a module as *not fault-prone* when actually, it is *fault-prone*. Table 10.1 summarizes our notation. In a standard development process, the expected proportion of *fault-prone* modules is π_{fp}, and similarly for *not fault-prone* modules, π_{nfp}. As explained in Table 10.1, the expected number of modules in various subsets are shown in Table 10.2, where in addition to totals, the rows signify actual classes of modules, and the columns correspond to classes predicted by a model.

In software engineering practice, the penalty for a Type II misclassification is often much more severe than for a Type I. A software enhancement technique, such as extra reviews, typically has modest direct cost per module. The cost of a Type I misclassification, C_I is the effort wasted on a not fault-prone module. On the other hand, the cost of a Type II misclassification, C_{II}, is the lost opportunity to correct faults early. The consequences of letting a fault go undetected prior to system release can be very expensive indeed. In this paper, we model the costs of misclassifying a module, C_I and C_{II}, as constants [12]. Future research will consider more sophisticated cost functions.

The expected cost of misclassification of one module, *ECM*, takes prior probabilities and costs of misclassification into account.

$$ECM = C_I \Pr(fp \mid nfp)\, \pi_{nfp} + C_{II} \Pr(nfp \mid fp)\, \pi_{fp} \qquad (10.11)$$

Hence, a classification rule that minimizes the expected cost of misclassification [27] is,

$$Class(\mathbf{x}_i) = \begin{cases} nfp & \text{if } \frac{f_{nfp}(\mathbf{x}_i)}{f_{fp}(\mathbf{x}_i)} \geq \left(\frac{C_{II}}{C_I}\right)\left(\frac{\pi_{fp}}{\pi_{nfp}}\right) \\ fp & \text{otherwise} \end{cases} \qquad (10.12)$$

There is a tradeoff between the Type I misclassification rate and the Type II misclassification rate. As one goes down, the other goes up. We have observed that it is often difficult to determine a practical balance for this tradeoff in terms of Type I and Type II misclassification rates. Therefore,

Table 10.1. Notation

G_{nfp}	the *not fault-prone* class (group) of modules
G_{fp}	the *fault-prone* class (group) of modules
k	index of classes
n	the number of modules in the software system under study
i	index of modules
$\mathbf{x}_i$	the vector of independent variables of the ith module
$Class(\mathbf{x}_i)$	the ith module's class, predicted by a model's classification rule
$f_k(\mathbf{x}_i)$	a likelihood function for the ith module's membership in the class k
$\hat{f}_k(\mathbf{x}_i)$	an estimate of $f_k(\mathbf{x}_i)$
π_k	the prior probability of membership in G_k
n_k	the expected number of modules in G_k
$\Pr(nfp \mid nfp)$	the probability that a model correctly classifies a module as in G_{nfp}
$\Pr(fp \mid fp)$	the probability that a model correctly classifies a module as in G_{fp}
$\Pr(fp \mid nfp)$	the probability that a model misclassifies a module as in G_{fp} which is actually in G_{nfp}, i.e., the Type I misclassification rate
$\Pr(nfp \mid fp)$	the probability that a model misclassifies a module as in G_{nfp} which is actually in G_{fp}, i.e., the Type II misclassification rate
$\hat{n}_{nfp,nfp}$	the expected number of modules that a model correctly classifies as in G_{nfp}
$\hat{n}_{fp,fp}$	the expected number of modules that a model correctly classifies as in G_{fp}
$\hat{n}_{nfp,fp}$	the expected number of modules that a model misclassifies as in G_{fp} which are actually in G_{nfp}
$\hat{n}_{fp,nfp}$	the expected number of modules that a model misclassifies as in G_{nfp} which are actually in G_{fp}
C_I	the cost of a Type I misclassification
C_{II}	the cost of a Type II misclassification

we translate these rates into measures of the *effectiveness* and *efficiency* of a model, which are more closely related to project management concerns.

Following the model's recommendation, the proportion of modules that are actually *fault-prone* and receive reliability enhancement is $\Pr(fp \mid fp)\pi_{fp}$. We define *effectiveness* as the proportion of *fault-prone* modules that received reliability enhancement treatment out of all the *fault-prone* modules.

$$effectiveness = \frac{\Pr(fp \mid fp)\pi_{fp}}{\pi_{fp}} = 1 - \Pr(nfp \mid fp) \qquad (10.13)$$

Table 10.2. Notation for Module Counts

Class	Model G_{nfp}	G_{fp}	Total
Actual			
G_{nfp}	$\hat{n}_{nfp,nfp}$	$\hat{n}_{nfp,fp}$	n_{nfp}
G_{fp}	$\hat{n}_{fp,nfp}$	$\hat{n}_{fp,fp}$	n_{fp}
Total			n

One can maximize *effectiveness* by minimizing the Type II misclassification rate, $\Pr(nfp \mid fp)$.

When one applies a reliability enhancement process to a *not fault-prone* module, s/he is actually wasting time, because the reliability of the *not fault-prone* module is already satisfactory. We define *efficiency* as the proportion of reliability enhancement effort that is not wasted. This is equivalent to the proportion of *fault-prone* modules that received reliability enhancement treatment out of all modules that received it.

$$efficiency = \frac{\Pr(fp \mid fp)\pi_{fp}}{\Pr(fp \mid nfp)\pi_{nfp} + \Pr(fp \mid fp)\pi_{fp}} \tag{10.14}$$

One can maximize *efficiency* by minimizing the Type I misclassification rate, $\Pr(fp \mid nfp)$.

Our goal is to give appropriate emphasis on *effectiveness* and *efficiency* based upon the needs of the project. The classification rule, mentioned in the following section (Section 10.3.2), enables a project to keep the best balance between the misclassification rates, and consequently, between *effectiveness* and *efficiency*.

10.3.2   Heuristic Method

Suppose there are n input-output pairs $(\mathbf{x}_i, Class(\mathbf{x}_i))$, $i = 1, 2, ..., n$ in the training data set for a m-input-single-output fuzzy rule-based system. Our goal is to derive a class number, i.e., 0 or 1, for each fuzzy if-then rule from the training data.

In order to determine the consequent $b_{p_1 \ldots p_m}$ of the fuzzy rule $R_{p_1 \ldots p_m}$ in Equation 10.6, we need to calculate the combination weight of the ith input-output pair $(\mathbf{x}_i, Class(\mathbf{x}_i))$ as,

$$W_{p_1 \ldots p_m}(\mathbf{x}_i) = \mu_{p_1 \ldots p_m}(\mathbf{x}_i) \tag{10.15}$$

We then apply the weight to the following heuristic method to determine a real number between 0 and 1,

$$b'_{p_1 \ldots p_m} = \sum_{i=1}^{n} W_{p_1 \ldots p_m}(\mathbf{x}_i) \cdot Class(\mathbf{x}_i) / \sum_{i=1}^{n} W_{p_1 \ldots p_m} \qquad (10.16)$$

and,

$$b_{p_1 \ldots p_m} = \begin{cases} nfp & \text{if } b'_{p_1 \ldots p_m} \geq \left(\frac{C_{II}}{C_I}\right)\left(\frac{\pi_{fp}}{\pi_{nfp}}\right) \\ fp & \text{otherwise} \end{cases} \qquad (10.17)$$

In Equation 10.16, the consequent real number $b'_{p_1 \ldots p_m}$ is determined as the weighted mean value of $Class(\mathbf{x}_i)$. A fuzzy rule can be generated from Equations 10.16 and (10.17) if there is at least one of the weights $W_{p_1 \ldots p_m}(\mathbf{x}_i)$ is not zero for $i = 1, 2, \ldots, n$. If all of the weights $W_{p_1 \ldots p_m}(\mathbf{x}_i)$ are zero, we are not able to generate the rule for that certain fuzzy subspace. In other words, there is no input-output pair $(\mathbf{x}_i, Class(\mathbf{x}_i))$ match with the fuzzy subspace $(A_{1p_1}, A_{2p_2}, \ldots, A_{mp_m})$.

10.3.3   Maximum Matched Method

The heuristic method actually calculates the comprehensive contribution of each input-output pair to the final consequent. This approach generates rules from one fuzzy subspace to another. Furthermore, we find it is difficult to determine how the individual rule affects the misclassification rate because the rules are derived from the average influence. In order to make it easy to modify the rules manually so as to reduce the cost of misclassifications, we need to observe how an individual rule contributes to the misclassification rate. The maximum matched method we propose here can address this problem.

We start from the input-output pair instead of the fuzzy subspace so that the performance of rules can be recorded. For a given $\mathbf{x}_i = (x_{i1}, x_{i2}, \ldots, x_{im})$, we can obtain its *maximum matched* fuzzy subspaces $(A_{1p_1}, A_{2p_2}, \ldots, A_{mp_m})$ as following.

$$\{(A_{1p_1}, A_{2p_2}, \ldots, A_{mp_m}) \mid \mu_{A_{1p_1}}(x_{i1}) = sup(\mu_{A_1}(x_{i1})), \ldots, \mu_{A_{mp_m}}(x_{im})$$
$$= sup(\mu_{A_m}(x_{im}))\} \qquad (10.18)$$

where $sup(\mu_{A_j}(x_{ij}))$ represents the supremum of $\mu_{A_j}(x_{ij})$ over the space of A_j. So the input $\mathbf{x}_i = (x_{i1}, x_{i2}, \ldots, x_{im})$ has the greatest influence on its maximum matched fuzzy subspaces. The output associated with this input will add a "credit" (fault-prone or not fault-prone) on the rule associated with this fuzzy subspace. We add two elements in the fuzzy subspace as

$$(A_{1p_1}, A_{2p_2}, \ldots, A_{mp_m}, Count(nfp), Count(fp)) \qquad (10.19)$$

We define the subspace described in Equation 10.19 as *extended fuzzy subspace*, where $Count(fp)$ and $Count(nfp)$ represent the number of fault-prone and not fault-prone modules which maximally match this fuzzy subspace over all the input-output pairs, respectively. We can derive the fuzzy rules from the extended fuzzy subspaces as following

$$
b_{p_1 \ldots p_m} = \begin{cases} nfp & \text{if } \text{Count}(nfp)/\text{Count}(fp) \geq \left(\frac{C_{II}}{C_I}\right)\left(\frac{\pi_{fp}}{\pi_{nfp}}\right) \\ fp & \text{otherwise} \end{cases} \tag{10.20}
$$

If $\text{Count}(fp) = \text{Count}(nfp) = 0$ for certain extended fuzzy subspace, we can not generate any fuzzy rule for this fuzzy subspace. It is easy to see how an individual rule affects the misclassification rate from an extended fuzzy subspace. For example, if we generate a rule from extended fuzzy space $(A_{11}, A_{21}, ..., A_{m1}, 25, 3)$ as,

Rule $R_{1\ldots1}$: If $(x_{i1}$ is $A_{11})$ ... and (If x_{im} is A_m) then $(y_i$ is $nfp)$ (10.21)

This rule will generate three Type II errors for the training data set. We say the *Type II misclassification count* for this rule is 3 . If we change the consequent of the rule to *fp*, it will generate 25 Type I errors and hence, the *Type I misclassification count* is 25. Manual change can be applied to the extended fuzzy subspace to modify the rule easily so that we can balance the Type I and Type II misclassification counts.

Theoretically, we can reduce the misclassification rate by dividing the input variables into more fuzzy sets which will increase the number of fuzzy subspace exponentially. However, if the number of fuzzy subspaces is too large, the amount of calculation will be enormous. For example, suppose we have 10 input variables and each of them is divided into 10 fuzzy sets, then the total number of subspaces of the 10 input variables will be 10^{10}. Such an approach will take a long time to generate rules and use them to obtain useful results. Another problem it may cause is overfitting. In the statistics literature, overfitting is often defined as *bias*, namely, the difference between a model's actual misclassification rate and its purported rate [28].

In order to increase the accuracy of the model while reducing the amount of subspaces and the chance of overfitting, we selectively divided the input variables into more fuzzy sets. Recall that in Section 10.2.1 we described that a partial partition of input variables has a better manifestation of data features. We utilize this partial partition before we apply the maximum matched method to generate rules. If the misclassification rate is higher than the expected rate, we will select the extended fuzzy subspaces with high misclassification counts, and try to divide the fuzzy sets in this fuzzy subspace into more fuzzy sets until the total misclassification rate decreases to an acceptable range.

10.4   Nonparametric Discriminant Analysis

Nonparametric discriminant analysis is a statistical technique for predicting the class (group), G_1 or G_2, of an observation, such as the ith software module, represented by its vector of independent variables, $\mathbf{x}_i$.

We use *stepwise discriminant analysis* at a significance level, α, to choose the independent variables in the nonparametric discriminant model [27]. An F test is used to test significance. Given a list of candidate variables, and beginning with no variables in the model, the variable not already in the model with the best significance level is added to the model, as long as its significance is higher than the threshold, α. Then the variable already in the model with the worst significance level is removed from the model as long as its significance is smaller than the threshold, α. These steps are repeated until no variable can be added or removed from the model.

We estimate a discriminant function based on the *fit* data set. Let $f_k(\mathbf{x}_i)$ be the multivariate probability density giving the likelihood that a module represented by $\mathbf{x}_i$ is in G_k. We predict the class membership of a module by comparing density functions f_1 and f_2 at $\mathbf{x}_i$. A large likelihood means the module is probably in that group. Since the density functions, f_k, are often non-normal distributions, we use "nonparametric" density estimation, and the multivariate kernel density estimation technique [27]. Let $\hat{f}_k(\mathbf{x}_i \mid \lambda)$ be an approximation of $f_k(\mathbf{x}_i)$, where λ is a smoothing parameter. The estimated density function is given by

$$\hat{f}_k(\mathbf{x}_i \mid \lambda) = \frac{1}{n_k} \sum_{l=1}^{n_k} K_k(\mathbf{x}_i \mid \mathbf{x}_{kl}, \lambda) \tag{10.22}$$

where n_k is the number of observations in group $G_k, k = 1, 2$, and the vector $\mathbf{x}_{kl}, l = 1, \ldots, n_k$ represents a module in group G_k, and $K_k(\mathbf{u} \mid \mathbf{v}, \lambda)$ is a multivariate kernel function on vector $\mathbf{u}$ with modes at $\mathbf{v}$. We use the normal kernel.

$$K_k(\mathbf{u} \mid \mathbf{v}, \lambda) = (2\pi\lambda^2)^{-n_k/2} \mid \mathbf{S}_k \mid^{-1/2} \exp(\,(-1/2\lambda^2)(\mathbf{u} - \mathbf{v})'\mathbf{S}_k^{-1}(\mathbf{u} - \mathbf{v})\,) \tag{10.23}$$

where, $\mathbf{S}_k$ is the covariance matrix for all observations in G_k, and $\mid \mathbf{S}_k \mid$ is its determinant.

We have observed that the Type I and Type II misclassification rates vary in opposite directions, within limits, as λ varies. We empirically choose a preferred value for λ based on misclassification rates of cross-validation [32] using the *fit* data set. We prefer approximately equal misclassification rates to the maximum extent feasible.

Let π_k be the prior probability of membership in G_k, which we usually choose to be the proportion of *fit* observations in G_k. A classification rule that minimizes the expected number of misclassification is given by the following,

$$Class(\mathbf{x}_i) = \begin{cases} G_1 & \text{if } \frac{\hat{f}_1(\mathbf{x}_i|\lambda)}{\hat{f}_2(\mathbf{x}_i|\lambda)} \geq \frac{\pi_2}{\pi_1} \\ G_2 & \text{otherwise} \end{cases} \tag{10.24}$$

10.5   Case-Based Reasoning

A case-based reasoning technique aims to find solutions to new problems based on past experience. The past experience is represented by "cases" in a "case library". The case library and the associated retrieval and decision rules constitute a CBR model. The past instances or cases are well-known project data from previously developed systems or projects, and contain all relevant information pertaining to each case. When associating with software quality modeling based on software metrics, a case in the case library is composed of a set of predictors or *independent* variables ($\mathbf{x}_i$), and a response or *dependent* variable (y_i).

In software quality classification modeling, the response variable is a *class membership*, i.e., either *fault-prone* or *not fault-prone*. The definition of whether a module is *fault-prone* or *not fault-prone* depends on the pre-set threshold value of the quality factor, which in our studies is the number of faults. Consequently, if the number of faults is fewer than the threshold value, the module is deemed as not fault-prone, and fault-prone otherwise. Suppose each case in the library has known attributes and class membership. Then, given a case with unknown class, we predict its class to be the same as the class of the most *similar* case(s) in the case library, where similarity is defined in terms of the case attributes.

A CBR classification model uses a *similarity function* to determine the most similar cases to the current case, from the case library (*fit* or *training* data). The function computes the distance d_{ij}, between the current case $\mathbf{x}_i$, and every other case $\mathbf{c}_j$ in the case library. The cases with the smallest possible distances are of primary interest, and the set of similar cases forms the set of *nearest neighbors*, N. Model parameter $\mathbf{n}_N$, represents the number of the best (most similar to current case) cases selected from N for case analysis and class estimation. $\mathbf{n}_N$ can be varied during model calibration to obtain different classification models. Once $\mathbf{n}_N$ is selected, a classification technique is used to classify the current case as either fault-prone or not fault-prone.

There are several types of similarity functions available including, *city block distance, euclidean distance*, and *mahalonobis distance*. Our previous research [20,22], indicated that the *mahalonobis* distance similarity function yielded better classification accuracy than the other two functions. Consequently, we use this distance function in our case studies. The *mahalonobis* distance similarity function is given:

$$d_{ij} = (\mathbf{c}_j - \mathbf{x}_i)' S^{-1} (\mathbf{c}_j - \mathbf{x}_i) \tag{10.25}$$

where, S is the variance-covariance matrix of the independent variables for the case library and S^{-1} is its inverse. Prime ($\prime$) indicates the transpose. Unlike the *euclidean* and *city block* distance functions, the *mahalonobis* distance function explicitly accounts for the correlation among the attributes, and does not require *standardization* or *normalization* of the independent variables [22].

A classification model may be sensitive to the ratio of the costs of the Type I and Type II misclassifications. This is because in practice actual costs of the two misclassifications (C_I and C_{II}) are not known during the modeling period. Practically speaking, the costs of the two misclassifications are not the same. Hence, the use of equal costs for C_I and C_{II} during model calibration is not realistic.

Moreover, other factors besides cost ratio may determine the best balance between the Type I and Type II error rates. For example, when the proportion of fault-prone modules (as compared to not fault-prone modules) is very small and Type II misclassifications have much more severe consequences than Type I, one may prefer equal misclassification rates. The cost ratio, $\frac{C_I}{C_{II}}$, can be varied during modeling to obtain a preferred balance [12] between the misclassification error rates.

In the context of CBR software classification models, we propose two methods of classification, i.e., *majority voting* and *data clustering*. It was indicated that the *data clustering* technique yielded better classification results. Consequently, we use it as the classification technique for our case studies. In the *data clustering* method, the case library is partitioned into two clusters, fault-prone (*fp*) and not fault-prone (*nfp*), according to the class of each case. For a currently unclassified case $\mathbf{x}_i$, let $d_{nfp}(\mathbf{x}_i)$ be the average distance to the not fault-prone nearest neighbor cases, and $d_{fp}(\mathbf{x}_i)$ be the average distance to the fault-prone nearest neighbor cases. The number of nearest neighbor cases to be used for analysis, can be varied (as a model parameter) during modeling.

Once the average distances to the nearest neighbor cases are computed, our *proposed* generalized classification rule for *data clustering* is then used to estimate the class, $Class(\mathbf{x}_i)$, of the unclassified case. The classification rule is given by:

$$Class(\mathbf{x}_i) = \begin{cases} nfp \text{ If } \frac{d_{fp}(\mathbf{x}_i)}{d_{nfp}(\mathbf{x}_i)} > \frac{C_I}{C_{II}} \\ fp \text{ otherwise} \end{cases} \tag{10.26}$$

A CBR software quality classification model therefore, constitutes of the following: $\frac{C_I}{C_{II}}$, the modeling cost ratio; $\mathbf{n}_N$, the number of nearest neighbor cases for analysis and class estimation; a *similarity function*, such as the mahalonobis distance, and a *classification method*, such as data clustering.

10.6   Network Communication System

10.6.1   System Description

We studied a network communication system (NT) written by professional programmers at a major corporation. This network computer application included numerous finite state machines and interfaces to other kinds of equipment. This case study did not include all of the modules in this application, we took a large sample of the modules for our analysis. The sample modules consist of about 1.3 million lines of code. In this case study we selected product metrics that can be collected during the detailed design phase. We chose call-graph metrics that can be collected as early as the high-level design phase. We also chose certain control flow-graph metrics that can be collected from the detailed design.

Table 10.3. System Profile for NT

Application	Telecommunications
Languate	Pascal-like
Lines of Code	1.3 million
Executable Statements	1.0 million
CFG Edges	364 thousand
Source Files	25 thousand
Functional Modules	2 thousand
Design Metrics	9
PCA Domain Metrics	3
Reuse Covariates	2
Quality Metric	Number of faults

Table 10.3 gives a profile of the metrics in this study. The quality metric, *Faults*, is the total number of faults discovered from coding through operations. Because of the module reuse, over half of the modules had no faults. Table 10.4 lists the product metrics that were used in this study. These metrics were derived from a module's call-graph or its control-flow graph.

We randomly split the data into a fit data set (1,320 modules) and a test data set (660 modules) to simulate two projects. Direct records of reuse history were not available. However, we derived reuse history information [19] from product metric "deltas". Delta variables were different for the current and prior release measurements, which enabled us to define categorical variables for reuse history. New modules had no prior release.

Table 10.4. Design product metrics for NT

Symbol	Title/Description
Call Graph Metrics	
MU	Modules used. The number of modules that this module uses directly, including itself
TC	Total calls to others. The number of calls to entry points in other procedures
UC	Unique calls to others. The number of unique entry points called by this module
Control Flow Graph Metrics	
IFTH	If-Then conditional arcs. The number of arcs that contain a predicate of a control structure, but are not loops
LP	Loops. The number of backward arcs
NL	Nesting level. The total nesting level of all arcs
SPC	Span of conditional arcs. The total number of arcs located within the span of conditional arcs
SPL	Span of loops. The number of vertices plus the number of arcs within loop control structure spans
VG	McCabe cyclomatic complexity [6]

Table 10.5. Domain Pattern for NT Software Metrics

Metric	PCA1	PCA2	PCA3
SPL	**0.901**	0.359	0.137
LP	**0.880**	0.370	0.134
SPC	**0.719**	0.545	0.316
NL	**0.683**	0.593	0.334
TC	0.359	**0.864**	0.216
UC	0.426	**0.830**	0.245
VG	0.597	**0.724**	0.309
IFTH	0.599	**0.681**	0.357
MU	0.177	0.265	**0.939**
Variance	3.630	3.410	1.460
% Variance	40.3%	37.9%	16.4%
Cumulative	40.3%	78.1%	94.4%

Stopping rule: at least 94% of variance

$$IsNew = \begin{cases} 1 \text{ All delta variables are missing} \\ 0 \text{ Otherwise} \end{cases} \qquad (10.27)$$

Those modules with unchanged measurements in any of the available metrics will be treated as reused modules.

$$IsChg = \begin{cases} 0 \text{ All delta variables are zero} \\ 1 \text{ Otherwise} \end{cases} \qquad (10.28)$$

Preexisting modules with some changed measurements were reused with modifications. *IsNew* and *IsChg* were treated as independent variables in our models.

The dependent variable was whether a module was fault-prone or not. We defined the not fault-prone group as those modules with fewer than five faults and the fault-prone group as those modules with five or more faults. About 12 percent of the modules in the fit data set were fault-prone. The proportion of modules considered fault-prone will vary with development efforts. We determined the threshold of five faults for this sample after discussing and consulting with engineers at the software development corporation. Published reports identify the proportion of high-risk modules to be anywhere between 5 and 20 percent.

Prepare Data using Principal Components Analysis Principal components analysis of the combined fit and test standardized design product metrics retained three components that accounted for 94.4% of the variance. Table 10.5 [18] shows the relationship between the original metrics and the domain metrics. Each table entry is the correlation between the metrics, the largest in each row is shown in bold.

The principal components analysis transformation was applied to the fit and test data sets. We can observe that *PCA1* is highly correlated with four metrics, including *SPL*, *LP*, *SPC*, and *NL*; *PCA2* is highly correlated with *TC*, *UC*, *VG*, and *IFTH*; *PCA3* with the metric *MU*. In this case study, the candidate independent variables were *PCA1, PCA2, PCA3, IsNew, and IsChg.*

10.6.2   Nonparametric Discriminant Analysis

We applied nonparametric discriminant analysis to the NT fit data set to estimate the discriminant function. We empirically selected a smoothing parameter, $\lambda = 0.05$, as discussed in Section 10.4. Table 10.6 shows the results when the model was applied to the test data set. The proportion of modules recommended for reliability enhancement was 31.4%. This model had useful accuracy: the Type I misclassification rate was 23.8%, the Type II

Table 10.6. NT Discriminant Analysis

Test data set
Number of Observations/Percent

	Model		
Class	*not f-p*	*f-p*	Total
Actual			
not f-p	442	138	580
	76.2%	23.8%	100.0%
f-p	11	69	80
	13.8%	86.2%	100.0%
Total	453	207	660
Percent	68.6%	31.4%	100.0%
Prior	87.9%	12.1%	

Overall misclassification: 22.6%

rate was 13.57%, and the overall rate was 22.6%. The misclassification rates translated into an effectiveness of 86.2% and an efficiency of 35.2%. In other words, 86.2% of the fault-prone modules were recommended for reliability enhancement, and 35.2% of the modules that received reliability enhancement would, in fact, be fault-prone.

10.6.3   Fuzzy Classification

We applied fuzzy classification to the NT data. The candidate independent variables were the same product metrics, as listed in Table 10.4. The dependent variable was whether a module was fault-prone or not. The same threshold was employed here to classify fault-prone or not fault-prone modules. Membership functions need to be defined before rules are generated. We divided the independent variables into fuzzy set according to the statistics on their distributions as follows:

- Membership functions for *PCA1* (Fig. 10.6):

$$
\begin{cases}
\mu_{\text{mf1}}(x) & = [-8, -4, -2.5] \\
\mu_{\text{mf2}}(x) & = [-4, -2.5, -1.25] \\
\mu_{\text{mf3}}(x) & = [-2.5, -1.25, -0.6] \\
\mu_{\text{mf4}}(x) & = [-1.25, -0.6, 0] \\
\mu_{\text{mf5}}(x) & = [-0.6, 0, 0.6] \\
\mu_{\text{mf6}}(x) & = [0, 0.6, 1.25] \\
\mu_{\text{mf7}}(x) & = [0.6, 1.25, 1.85] \\
\mu_{\text{mf8}}(x) & = [1.25, 1.85, 2.5] \\
\mu_{\text{mf9}}(x) & = [1.85, 2.5, 3.75] \\
\mu_{\text{mf10}}(x) & = [2.5, 3.75, 5] \\
\mu_{\text{mf11}}(x) & = [3.75, 5, 26]
\end{cases}
\tag{10.29}
$$

- Membership functions for *PCA2* (Fig. 10.7):

$$
\begin{cases}
\mu_{\text{mf1}}(x) & = [-6, -4, -2.5] \\
\mu_{\text{mf2}}(x) & = [-4, -2.5, -1.25] \\
\mu_{\text{mf3}}(x) & = [-2.5, -1.25, -0.6] \\
\mu_{\text{mf4}}(x) & = [-1.25, -0.6, 0] \\
\mu_{\text{mf5}}(x) & = [-0.6, 0, 0.6] \\
\mu_{\text{mf6}}(x) & = [0, 0.6, 1.25] \\
\mu_{\text{mf7}}(x) & = [0.6, 1.25, 1.85] \\
\mu_{\text{mf8}}(x) & = [1.25, 1.85, 2.5] \\
\mu_{\text{mf9}}(x) & = [1.85, 2.5, 3.75] \\
\mu_{\text{mf10}}(x) & = [2.5, 3.75, 5] \\
\mu_{\text{mf11}}(x) & = [3.75, 5, 26]
\end{cases}
\tag{10.30}
$$

- Membership functions for *PCA3* (Fig. 10.8):

$$
\begin{cases}
\mu_{\text{mf1}}(x) & = [-6, -4, -2.5] \\
\mu_{\text{mf2}}(x) & = [-4, -2.5, -1.25] \\
\mu_{\text{mf3}}(x) & = [-2.5, -1.25, -0.6] \\
\mu_{\text{mf4}}(x) & = [-1.25, -0.6, 0] \\
\mu_{\text{mf5}}(x) & = [-0.6, 0, 0.6] \\
\mu_{\text{mf6}}(x) & = [0, 0.6, 1.25] \\
\mu_{\text{mf7}}(x) & = [0.6, 1.25, 1.85] \\
\mu_{\text{mf8}}(x) & = [1.25, 1.85, 2.5] \\
\mu_{\text{mf9}}(x) & = [1.85, 2.5, 3.75] \\
\mu_{\text{mf10}}(x) & = [2.5, 3.75, 5] \\
\mu_{\text{mf11}}(x) & = [3.75, 5, 26]
\end{cases}
\tag{10.31}
$$

- Membership functions for *IsChg* (Fig. 10.9):

$$
\begin{cases}
\mu_{\text{mf1}}(x) = [-0.5, 0, 0.5] \\
\mu_{\text{mf2}}(x) = [0.5, 1, 1.5]
\end{cases}
\tag{10.32}
$$

- Membership functions for *IsNew* (Fig. 10.10):

$$\begin{cases} \mu_{\mathrm{mf1}}(x) = [-0.5, 0, 0.5] \\ \mu_{\mathrm{mf2}}(x) = [0.5, 1, 1.5] \end{cases} \tag{10.33}$$

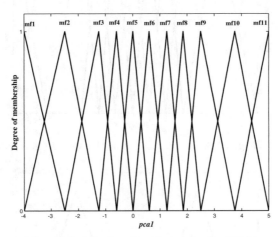

Fig. 10.6. Membership Functions for *PCA1*

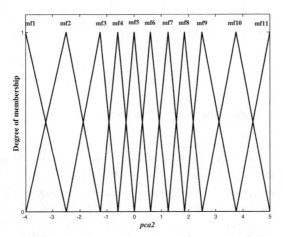

Fig. 10.7. Membership Functions for *PCA2*

We employed the maximum matched method to generate the rules based on the membership functions defined above, and we empirically choose $\left(\frac{C_{II}}{C_I}\right)\left(\frac{\pi_{fp}}{\pi_{nfp}}\right) = 3.0$. Table 10.7 shows the rules generated by the maximum matched method for the NT data.

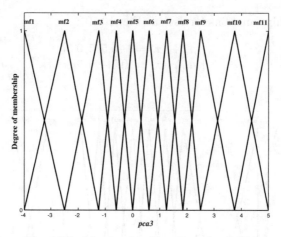

Fig. 10.8. Membership Functions for *PCA3*

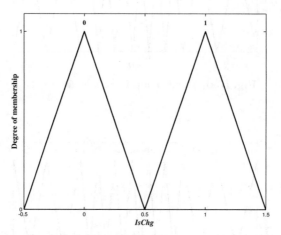

Fig. 10.9. Membership Functions for *IsChg*

Table 10.8 shows the results obtained after the model was applied to the test data set. The proportion of modules recommended for reliability enhancement was 25.6%. This model had better accuracy: the Type I misclassification rate was 16.5%, and the Type II rate was 7.8%. The overall rate was 12.8%. The misclassification rates translated into an *effectiveness* of 91.3% and an *efficiency* of 44.0%. In other words, 91.3% of the *fault-prone* modules were recommended for reliability enhancement, and 44.0% of the modules that received reliability enhancement would, in fact, be *fault-prone*.

Table 10.9 shows that fuzzy classification significantly improves the classification quality. The Type I misclassification rate decreased by 7.3% and Type II by 5.1% and overall misclassification rate by 7.0%. The effectiveness and efficiency increased by 5.1% and 8.8%, respectively.

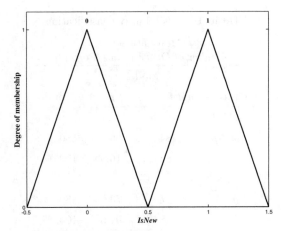

Fig. 10.10. Membership Functions for *IsNew*

Table 10.7. Snapshot of Fuzzy Rules for NT data

	Input				Output
PCA1	*PCA2*	*PCA3*	*IsChg*	*IsNew*	
mf9	mf10	mf6	0	1	*nfp*
mf9	mf11	mf5	0	1	*nfp*
mf10	mf3	mf5	0	1	*nfp*
mf10	mf9	mf3	0	1	*nfp*
mf11	mf6	mf3	0	1	*nfp*
mf11	mf10	mf3	0	1	*nfp*
mf1	mf11	mf9	0	1	*fp*
mf2	mf7	mf4	0	1	*fp*
mf2	mf9	mf10	0	1	*fp*
mf2	mf10	mf3	0	1	*fp*
mf3	mf2	mf9	0	1	*fp*
mf3	mf4	mf9	0	1	*fp*
mf3	mf5	mf5	0	1	*fp*

Table 10.8. NT Fuzzy Classification

test data set
Number of Observations/Percent

	Model		
Class	*not f-p*	*f-p*	Total
Actual			
not f-p	484	96	580
	83.5%	16.5%	100.0%
f-p	7	73	80
	8.7%	91.3%	100.0%
Total	491	169	660
Percent	74.4%	25.6%	100.0%
Prior	87.9%	12.1%	

Overall misclassification: 15.6%

Table 10.9. NT Comparison Result

	Model	
	Discriminant	Fuzzy Classification
Type I Error	23.8%	16.5%
Type II Error	13.8%	8.7%
Overall Error	22.6%	15.6%
Effectiveness	86.2%	91.3%
Efficiency	35.2%	44.0%

10.7   An Ada System

10.7.1   System Description

We studied a large command, control, and communications system implemented in Ada, which we call CCCS. CCCS was developed at a large organization by professionals using the procedural programming paradigm. The providers of the CCCS data set collected faults from problem tracking reports generated during the system integration and test phase. Generally, a module was an Ada package, that consists of one or more procedures. The top 20% of the modules contained 82.2% of the faults, and the top 5% of the modules contained 40.2% of faults. Therefore, it is of great value to identify the modules with the most faults before the testing phase. 52% of the modules had no faults, and over 75% modules had two or fewer faults. The software product metrics used in the CCCS system are listed in Table 10.10.

Table 10.10. Software Product Metrics for CCCS

Symbol	Description
η_1	Number of unique operators [6].
N_1	Total number of operators [6]
η_2	Number of unique operands [6]
N_2	Total number of operands [6]
$V(G)$	McCabe's cyclomatic complexity
N_L	Number of logical operators
LOC	Lines of code
$ELOC$	Executable lines of code

Table 10.11. Distribution of the Number of fault in CCCS

Statistic	fit	test
Modules	188	94
Min	0	0
Max	29	42
Mean	2.27	2.56
Std Dev	4.65	5.88

We analyzed all 282 modules measured by the developers. Applying data splitting, we randomly partitioned this data into two subsets, two thirds of the observations for a fit data set, the remaining third for a test data set. Table 10.11 gives summary statistics on the number of faults of the fit and test data sets. Table 10.12 lists descriptive statistics for the fit data set of the software metrics.

10.7.2   Nonparametric Discriminant Analysis

We applied nonparametric discriminant analysis to the CCCS data [27]. The dependent variable was whether a module was fault-prone or not. Fault-prone modules had four or more faults. The candidate independent variables were the same product metrics, as listed in Table 10.10. Stepwise discriminant analysis selected η_2, $V(G)$, η_1, $ELOC$, and N_2 at the 15% significance level.

Table 10.12. Descriptive statistics of CCCS software metrics

Metric	Mean	Std Dev	Min	Median	Max
η_1	2.7819	15.6150	4.0000	21.0000	85.0000
η_2	110.9000	152.8000	2.0000	61.0000	1124.0000
N_1	589.3000	1069.5000	6.0000	187.0000	8606.0000
N_2	499.4000	941.9000	2.0000	164.0000	7736.0000
$V(G)$	29.6755	68.3805	1.0000	8.0000	614.0000
N_L	3.1915	7.9957	0.0000	0.0000	18.0000
LOC	808.9000	1142.4000	19.0000	389.5000	9163.0000
$ELOC$	108.4000	182.9000	3.0000	42.5000	1412.0000

The significance of each parameter was $\alpha \leq 0.042$. We empirically selected a smoothing parameter, $\lambda = 0.20$, as discussed in Section 10.4. Table 10.13 shows the results when the model was applied to the *test* data set. The proportion of modules recommended for reliability enhancement was 33.0%. This model had useful accuracy: the Type I misclassification rate was 20.0%, the Type II rate was 15.8%, and the overall rate was 19.2%. The misclassification rates translated into an *effectiveness* of 84.2% and *efficiency* of 50.0%. In other words, 84.2% of the *fault-prone* modules were recommended for reliability enhancement, and half of the modules given reliability enhancement would, in fact, be *fault-prone*.

10.7.3 Fuzzy Classification

We applied fuzzy classification to CCCS data. We experimentally selected η_2, $V(G)$, N_L, and η_1 as independent variables, and the dependent variable was whether a module was fault-prone or not. The same threshold was employed here to classify fault-prone or not fault-prone modules. Membership functions should be defined before we generate rules. We divided the independent variables into fuzzy sets according to the statistics on their distributions as following.

- Membership functions for η_1 (Fig. 10.11):

$$
\begin{cases}
\mu_{\text{mf1}}(x) = e^{\frac{-(x-3.36)^2}{23.94}} \\
\mu_{\text{mf2}}(x) = e^{\frac{-(x-12.00)^2}{3.07}} \\
\mu_{\text{mf3}}(x) = e^{\frac{-(x-15.90)^2}{3.03}} \\
\mu_{\text{mf4}}(x) = e^{\frac{-(x-21.4)^2}{13.01}} \\
\mu_{\text{mf5}}(x) = e^{\frac{-(x-34.6)^2}{24.50}} \\
\mu_{\text{mf6}}(x) = e^{\frac{-(x-69.1)^2}{499.28}}
\end{cases}
\tag{10.34}
$$

Table 10.13. CCCS Discriminant Analysis

test data set

Number of Observations/Percent

		Model	
Class	*not f-p*	*f-p*	Total
Actual			
not f-p	60	15	75
	80.0%	20.0%	100.0%
f-p	3	16	19
	15.8%	84.2%	100.0%
Total	63	31	94
Percent	67.0%	33.0%	100.0%
Prior	80.8%	19.2%	

Overall misclassification: 19.2%

- Membership functions for η_2 (Fig. 10.12):

$$\begin{cases} \mu_{\text{mf1}}(x) = \frac{1}{1+|\frac{x}{10.3}|^{5.00}} \\ \mu_{\text{mf2}}(x) = \frac{1}{1+|\frac{x-19.1}{6.3}|^{5.00}} \\ \mu_{\text{mf3}}(x) = \frac{1}{1+|\frac{x-56.3}{20.7}|^{4.68}} \\ \mu_{\text{mf4}}(x) = \frac{1}{1+|\frac{x-104}{19.8}|^{7.64}} \\ \mu_{\text{mf5}}(x) = \frac{1}{1+|\frac{x-152}{17.1}|^{5.00}} \\ \mu_{\text{mf6}}(x) = [174, 185, 426, 429] \end{cases} \tag{10.35}$$

- Membership functions for N_L (Fig. 10.13):

$$\begin{cases} \mu_{\text{mf1}}(x) = e^{\frac{-(x-1.21)^2}{8.08}} \\ \mu_{\text{mf2}}(x) = e^{\frac{-(x-8.38)^2}{4.74}} \\ \mu_{\text{mf3}}(x) = e^{\frac{-(x-13.00)^2}{3.59}} \\ \mu_{\text{mf4}}(x) = e^{\frac{-(x-20.50)^2}{8.82}} \\ \mu_{\text{mf5}}(x) = [24.6, 27.1, 56, 58] \end{cases} \tag{10.36}$$

- Membership functions for $V(G)$ (Fig. 10.14):

$$\begin{cases} \mu_{\text{mf1}}(x) = [-6.66, 0, 3.33] \\ \mu_{\text{mf2}}(x) = [2.43, 4.34, 5.76] \\ \mu_{\text{mf3}}(x) = [3.92, 9.90, 15.10] \\ \mu_{\text{mf4}}(x) = [13.8, 17.8, 20.2, 35.8] \end{cases} \tag{10.37}$$

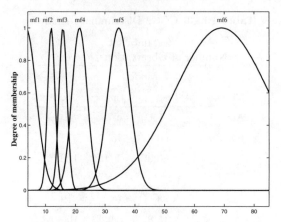

Fig. 10.11. Membership Functions for η_1

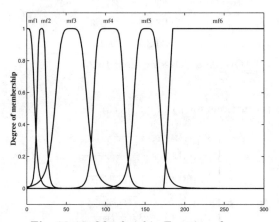

Fig. 10.12. Membership Functions for η_2

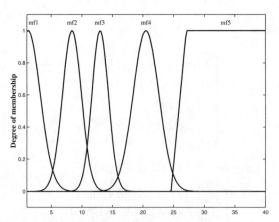

Fig. 10.13. Membership Functions for N_L

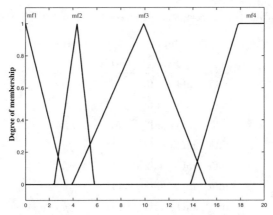

Fig. 10.14. Membership Functions for $V(G)$

We employed the maximum matched method to generate the rules based on the membership functions defined above. We experimentally choose $\left(\frac{C_{II}}{C_I} \right) \left(\frac{\pi_{fp}}{\pi_{nfp}} \right) = 3.0$. Table 10.14 shows the rules generated by the maximum matched method for the CCCS data.

Table 10.15 shows the results obtained after the model was applied to the test data set. The proportion of modules recommended for reliability enhancement was 28.7%. This model had better accuracy: the Type I misclassification rate was 13.3%, the Type II rate was 10.5%, and the overall rate was 12.8%. The misclassification rates translated into an effectiveness of 89.5% and efficiency of 61.5%. In other words, about 90% of the fault-prone modules were recommended for reliability enhancement, and around 60% of the modules given reliability enhancement would, in fact, be fault-prone.

Table 10.16 shows that fuzzy classification significantly improves the classification quality. The Type I misclassification rate decreased by 6.7% and Type II by 5.3% and overall misclassification rate by 6.4%. The effectiveness and efficiency increased by 5.3% and 11.5%, respectively.

10.8   A Very Large Window Application System

We also applied the case based reasoning (CBR) [22] and fuzzy classification on the VLWA data, respectively. VLWA is a Very Large Windows-based software Applications which were written in a high-level language with over 1400 source code files combined, and had over 27.5 million lines of code overall. Table 10.17 lists the software product metrics used in this study. Upon pre-processing and cleaning the data [22], the data set had 1211 program modules. We randomly split the available data into the fit and test data sets. The fit data set contains 807 modules and the test data set contains 404 modules, respectively. The dependent variables is the number of faults in

Table 10.14. Snapshot of Fuzzy Rules for CCCS data

	Input			Output
η_1	η_2	N_L	$V(G)$	
mf5	mf3	mf5	mf1	*nfp*
mf5	mf3	mf5	mf2	*nfp*
mf5	mf3	mf5	mf3	*nfp*
mf5	mf4	mf2	mf3	*fp*
mf5	mf4	mf2	mf1	*nfp*
mf5	mf4	mf3	mf1	*fp*
mf5	mf4	mf4	mf1	*fp*
mf5	mf4	mf5	mf2	*fp*
mf5	mf4	mf5	mf3	*fp*
mf5	mf4	mf5	mf1	*nfp*
mf5	mf5	mf4	mf1	*fp*
mf5	mf5	mf5	mf1	*fp*
mf5	mf5	mf5	mf2	*fp*
mf5	mf5	mf5	mf3	*fp*
mf5	mf6	mf4	mf1	*nfp*

Table 10.15. CCCS Fuzzy Classification

test data set
Number of Observations/Percent

	Model		
Class	*not f-p*	*f-p*	Total
Actual			
not f-p	65	10	75
	86.7%	13.3%	100.0%
f-p	2	17	19
	10.5%	89.5%	100.0%
Total	67	27	94
Percent	71.3%	28.7%	100.0%
Prior	80.8%	19.2%	

Overall misclassification: 12.8%

Table 10.16. Comparison Result

	Model	
	Discriminant	Fuzzy Classification
Type I Error	20.0%	13.3%
Type II Error	15.8%	10.5%
Overall Error	19.2%	12.8%
Effectiveness	84.2%	89.5%
Efficiency	50.0%	61.5%

every software module, and we choose a threshold of 2 to classify the software modules as fault-prone or not fault-prone. This implies that if a module had fewer than 2 faults then it was classified as not fault-prone, and fault-prone otherwise. The following two sections show our empirical results.

Table 10.17. Product and Process Metrics for VLWA

Symbol	Title/Description
Product Metrics	
LOCB	Number of lines of code for the source file prior to coding phase
LOCT	Number of lines of code for the source file prior to system test release
LOCA	Number of lines of commented code for the source file prior to coding phase
LOCS	Number of lines of commented code for the source file prior to system test release
Process Metrics	
NUMI	Number of times the source file was inspected prior to system test release

10.8.1   Case Based Reasoning

We calibrated classification models for the VLWA fit data set using a case base reasoning approach. The candidate independent variables are listed in Table 10.17. Table 10.18 shows the results when the model was applied to the test data set. This model had useful accuracy: the Type I misclassification rate was 15.7%, the Type II rate was 13.4%, and the overall rate was 15.4%. The misclassification rates translated into an effectiveness of 84.3% and an efficiency of 52.3%. In other words, 84.3% of the fault-prone modules were recommended for reliability enhancement, and 52.3% of the modules that received reliability enhancement would, in fact, be fault-prone.

Table 10.18. VLWA Case Based Reasoning

test data set

Number of Observations/Percent

	Model		
Class	*not f-p*	*f-p*	Total
Actual			
not f-p	284	53	337
	84.3%	15.7%	100.0%
f-p	9	58	67
	13.4%	86.6%	100.0%
Total	293	111	404
Percent	72.5%	27.5%	100.0%

Overall misclassification: 15.4%

10.8.2 Fuzzy Classification

We applied fuzzy classification to the VLWA data. The candidate independent variables are listed in Table 10.17. The dependent variable was whether a module was fault-prone or not. The same threshold was employed here to classify fault-prone or not fault-prone. Membership functions need to be defined before rules are generated. We divided the independent variables into fuzzy sets according to the statistics on their distributions as follows:

- Membership functions for $LOCB$ (Fig. 10.15):

$$\begin{cases} \mu_{\mathrm{mf1}}(x) = e^{\frac{-x^2}{0.26}} \\ \mu_{\mathrm{mf2}}(x) = e^{\frac{-(x-1.17)^2}{0.13}} \\ \mu_{\mathrm{mf3}}(x) = e^{\frac{-(x-2.00)^2}{0.09}} \\ \mu_{\mathrm{mf4}}(x) = e^{\frac{-(x-3.00)^2}{0.10}} \\ \mu_{\mathrm{mf5}}(x) = [3.32, 3.93, 70, 75] \end{cases} \quad (10.38)$$

- Membership functions for $LOCT$ (Fig. 10.16):

$$\begin{cases} \mu_{\mathrm{mf1}}(x) = e^{\frac{-x^2}{53.45}} \\ \mu_{\mathrm{mf2}}(x) = e^{\frac{-(x-13.20)^2}{8.00}} \\ \mu_{\mathrm{mf3}}(x) = e^{\frac{-(x-23.50)^2}{16.59}} \\ \mu_{\mathrm{mf4}}(x) = e^{\frac{-(x-37.00)^2}{20.87}} \\ \mu_{\mathrm{mf5}}(x) = [39.94, 47.56, 2100, 2135] \end{cases} \quad (10.39)$$

- Membership functions for *LOCA* (Fig. 10.17):

$$\begin{cases} \mu_{\text{mf1}}(x) = e^{\frac{-x^2}{67.98}} \\ \mu_{\text{mf2}}(x) = e^{\frac{-(x-17.90)^2}{5.58}} \\ \mu_{\text{mf3}}(x) = e^{\frac{-(x-36.80)^2}{74.18}} \\ \mu_{\text{mf4}}(x) = e^{\frac{-(x-70.50)^2}{269.12}} \\ \mu_{\text{mf5}}(x) = [81.1, 92.8, 4020, 4030] \end{cases} \tag{10.40}$$

- Membership functions for *LOCS* (Fig. 10.18):

$$\begin{cases} \mu_{\text{mf1}}(x) = e^{\frac{-x^2}{87.91}} \\ \mu_{\text{mf2}}(x) = e^{\frac{-(x-19.90)^2}{21.91}} \\ \mu_{\text{mf3}}(x) = e^{\frac{-(x-27.90)^2}{4.09}} \\ \mu_{\text{mf4}}(x) = e^{\frac{-(x-33.00)^2}{4.56}} \\ \mu_{\text{mf5}}(x) = [35.4, 41.3, 800, 810] \end{cases} \tag{10.41}$$

- Membership functions for *NUMI* (Fig. 10.19):

$$\begin{cases} \mu_{\text{mf1}}(x) = e^{\frac{-x^2}{148.26}} \\ \mu_{\text{mf2}}(x) = e^{\frac{-(x-22.80)^2}{14.36}} \\ \mu_{\text{mf3}}(x) = e^{\frac{-(x-29.10)^2}{3.49}} \\ \mu_{\text{mf4}}(x) = e^{\frac{-(x-41.30)^2}{27.83}} \\ \mu_{\text{mf5}}(x) = [46.5, 54.1, 3950, 3960] \end{cases} \tag{10.42}$$

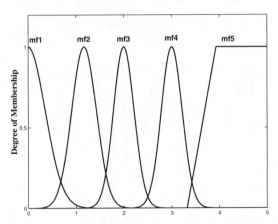

Fig. 10.15. Membership Functions for *LOCB*

We employed the maximum matched method to generate the rules based on the membership functions defined above. We empirically choose

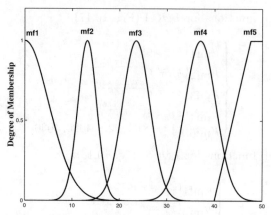

Fig. 10.16. Membership Functions for *LOCT*

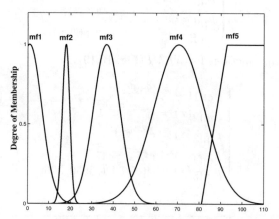

Fig. 10.17. Membership Functions for *LOCA*

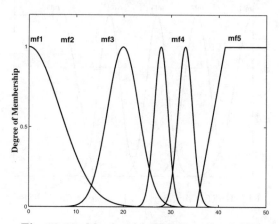

Fig. 10.18. Membership Functions for *LOCS*

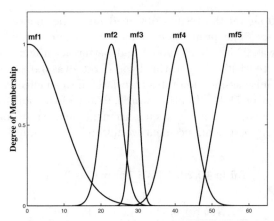

Fig. 10.19. Membership Functions for *NUMI*

$\left(\frac{C_{II}}{C_I}\right)\left(\frac{\pi_{fp}}{\pi_{nfp}}\right) = 3.0$, and Table 10.19 shows the rules generated for the VLWA data.

Table 10.19. Snapshot of Fuzzy Rules for VLWA data

		Input			Output
LOCB	*LOCT*	*LOCA*	*LOCS*	*NUMI*	
mf5	mf4	mf5	mf2	mf5	*nfp*
mf5	mf4	mf5	mf3	mf4	*nfp*
mf5	mf5	mf5	mf4	mf4	*nfp*
mf2	mf4	mf3	mf4	mf2	*nfp*
mf5	mf1	mf4	mf1	mf1	*nfp*
mf3	mf4	mf3	mf4	mf2	*nfp*
mf1	mf5	mf4	mf4	mf4	*nfp*
mf1	mf3	mf2	mf2	mf2	*nfp*
mf1	mf1	mf1	mf2	mf1	*fp*
mf1	mf1	mf2	mf1	mf2	*fp*
mf1	mf1	mf4	mf1	mf1	*fp*
mf1	mf3	mf3	mf3	mf2	*fp*
mf1	mf3	mf4	mf3	mf4	*fp*
mf1	mf3	mf4	mf3	m5	*fp*
mf1	mf4	mf3	mf2	mf3	*fp*

Table 10.20 shows the results obtained after the model was applied to the test data set. The proportion of modules recommended for reliability enhancement was 22.5%. This model had better accuracy: the Type I misclassification rate was 7.7%, and the Type II rate was 3.0%. The overall rate was 6.9%. The misclassification rates translated into an effectiveness of 97.0% and an efficiency of 71.4%. In other words, 97.0% of the fault-prone modules were recommended for reliability enhancement, and 71.4% of the modules that received reliability enhancement would, in fact, be fault-prone.

Table 10.20. VLWA Fuzzy Classification

test data set
Number of Observations/Percent

Class	Model not f-p	f-p	Total
Actual			
not f-p	311	26	337
	92.3%	7.7%	100.0%
f-p	2	65	67
	3.0%	97.0%	100.0%
Total	313	91	404
Percent	77.5%	22.5%	100.0%

Overall misclassification: 6.9%

Table 10.21. VLWA Comparison Result

	Model	
	CBR	Fuzzy Classification
Type I Error	15.7%	7.7%
Type II Error	13.4%	3.0%
Overall Error	15.4%	6.9%
Effectiveness	84.3%	97.0%
Efficiency	52.3%	71.4%

Table 10.21 shows that fuzzy classification significantly improves the classification quality. The Type I misclassification rate decreased by 8.0% and

the Type II by 10.4% and the overall misclassification rate by 8.5%. The effectiveness and efficiency increased by 12.7% and 19.1%, respectively.

10.9   Conclusions

Rather than predicting the number of faults, classification models serve quite well when only early identification of the most troublesome modules is needed. Software quality models are usually based on the software metrics collected at the end of the coding phase. The contribution of this paper is the introduction of the RBFC algorithm, in software reliability modeling. With RBFC, we can not only improve the robustness of the software quality model but can also increase its classification accuracy.

The rules of the RBFC were generated from the numerical data automatically in such a way so as to reduce the cost of misclassifications. We proposed a new algorithm called *maximum matched method* to extract the rules. With *extended fuzzy subspace*, practitioners are able to change the generated rules manually to balance the Type I and Type II misclassification rates.

In our case studies, we applied *maximum matched method* to extract rules from numerical input data, and RBFC used the generated rules based on software metrics to classify modules as either *fault-prone* or *not fault-prone*. The results of two of the case studies were compared with the nonparametric discriminant analysis classification models. The third case study was used to compare classification models built using the proposed RBFC method and case-based reasoning. It is observed that the RBFC models demonstrated very good classification accuracy. This gives management more flexible reliability enhancement strategies than discriminant analysis and case-based reasoning classification models. We also applied RBFC to other case studies, and similar results were obtained as well [32].

Future research will investigate more sophisticated analysis techniques, such as neural network, genetic programming for rule extraction. Approaches to reduce the size of the rule base may also be investigated. For example, one could merge two or more rules into one by using genetic programming.

10.10   Acknowledgements

This work was supported in part by Cooperative Agreement NCC 2-1141 from NASA Ames Research Center, Software Technology Division (Independent Verification and Validation Facility). We thank Naeem Seliya and Erik Geleyn for their useful suggestions and assistance with editorial reviews. The findings and opinions in this paper belong solely to the authors and are not necessarily those of the sponsors, or collaborators.

References

1. L. C. Briand, V. R. Basili, and C. J. Hetmanski. (1993) Developing interpretable models with optimized set reduction for identifying high-risk of software components. *IEEE Transaction on Software Engineering*, 19(11):1028–1044
2. K. H. Chen, H. L. Chen, and H. M. Lee. (1997) A multiclass neural network classifier with fuzzy teaching inputs. *Fuzzy Sets and Systems*, 91:15–35
3. C. Ebert. (1993) Rule-based fuzzy classification for software quality control. *Fuzzy Sets and Systems*, 63:349–358
4. C. Ebert. (1996) Classification techniques for metric-based software development. *Software Quality Journal*, 5:255–272
5. N. E. Fenton and S. L. Pfleeger. (1996) *Software Metrics: A Rigorous and Practical Approach*. PWS Publishing Company, Boston
6. M. H. Halstead. (1997) *Elements of Software Science*. Elsevier North-Holland, New York
7. H. Ichihashi and T. Watanabe. (1990) Learning control system by a simplified fuzzy reasoning model. *Proceeding: Information Processing and Management of Uncertainty*, pages 417–419
8. H. Ishibuchi, K. Nozaki, H. Tanaka, Y. Hosaka, and M. Matsuda. (1994) Empirical study on learning in fuzzy systems by rice taste analysis. *Fuzzy Sets and Systems*, 64:129–144
9. J.-S. R. Jang, C.-T. Sun, and E. Mizutani. (1997) *Neuro-Fuzzy And Software Computing: A Computational Approach to Learning and Machine Intelligence*. Prentice Hall, Upper Saddle River, NJ
10. W. D. Jones, J. P. Hudepohl, T. M. Khoshgoftaar, and E. B. Allen. (2002) Application of a usage profile in software quality models. *Science of Computer Programming*, In press.
11. C. L. Karr and E. J. Gentry. (1993) Fuzzy control of pH using genetic algorithms. *IEEE Transcation on Fuzzy Systems*, 1:46–53
12. T. M. Khoshgoftaar and E. B. Allen. (1998) Classification of fault-prone software modules: Prior probabilities, costs, and model evaluation. *Empirical Software Engineering: An International Journal*, 3(3):275–298
13. T. M. Khoshgoftaar and E. B. Allen. (1999) A comparative study of ordering and classification of fault-prone software modules. *Empirical Software Engineering: An International Journal*, 4:159–186
14. T. M. Khoshgoftaar and E. B. Allen. (1999) Predicting fault-prone software modules in embedded systems with classification trees. In *Proceedings: Fourth IEEE International Symposium on High-Assurance Systems Engineering*, pages 105–112, Washington, DC USA, Nov. IEEE Computer Society.
15. T. M. Khoshgoftaar and E. B. Allen. (2001) Modeling software quality with classification trees. In H. Pham, editor, *Recent Advances in Reliability and Quality Engineering*, chapter 15, pages 247–270. World Scientific, Singapore
16. T. M. Khoshgoftaar, E. B. Allen, N. Goel, A. Nandi, and J. McMullan. (1996) Detection of software modules with high debug code churn in a very large legacy system. In *Proceedings of the Seventh International Symposium on Software Reliability Engineering*, pages 364–371, White Plains, NY, Oct. IEEE Computer Society.
17. T. M. Khoshgoftaar, E. B. Allen, W. D. Jones, and J. P. Hudepohl. (2000) Accuracy of software quality models over multiple releases. *Annals of Software Engineering*, 9:103–116

18. T. M. Khoshgoftaar, E. B. Allen, K. S. Kalaichelvan, and N. Goel. (1996) Early quality prediction: A case study in telecommunications. *IEEE Software*, 13(1):65–71

19. T. M. Khoshgoftaar, E. B. Allen, K. S. Kalaichelvan, and N. Goel. (1996) The impact of software evolution and reuse on software quality. *Empirical Software Engineering: An International Journal*, 1(1):31–44

20. T. M. Khoshgoftaar, N. Sundaresh, and Z. Xu. (2000) Prediction of software faults using case based reasoning. Technical Report TR-CSE-01-29, Florida Atlantic University, Boca Raton, Florida USA

21. J. R. Koza. (1996) *Genetic Programming: on the Programming of Computers by Means of Natural Selection*. MIT Press, Cambridge, Massachusetts

22. L. Lim. (2001) Developing accurate software quality models using a faster, easier, and cheaper method. Master's thesis, Florida Atlantic University, Boca Raton, Florida USA, Advised by Taghi M. Khoshgoftaar.

23. E. H. Mamdani and S. Assilian. (1975) An experiment in linguistic synthesis with a fuzzy logic controller. *International Journal of Man-Machine Studies*, 7(1):1–13

24. D. Nauck and R. Kruse. (1997) A neuro-fuzzy method to learn fuzzy classifation rules from data. *Fuzzy Sets and Systems*, 89:277–288

25. H. Nomura, I. Hayashi, and N. Wakami. (1992) A learning method of fuzzy inference rules by descent method. *Proceeding: FUZZ-IEEE'92*, pages 203–210

26. T. Pfeufer and M. Ayoubi. (1997) Application of a hybrid neuro-fuzzy system to the fault diagnosis of an automotive electromechanical actuator. *Fuzzy Sets and Systems*, 89:351–360

27. G. A. F. Seber. (1984) *Multivariate Observations*. John Wiley & Sons, New York

28. M. Stone and J. Rasp. (1993) The assessment of predictive accuracy and model overfitting: An alternative approach. *Journal of Business Finance and Accounting*, 20(1):125–131

29. M. Sugeno and G. T. Kang. (1988) Structure identification of fuzzy model. *Fuzzy Sets and Systems*, 28:15–33

30. H. Takagi and M. Sugeno. (1985) Fuzzy identification of systems and its applications to modeling and control. *IEEE Transaction on Systems, Man, Cybernet*, 15:116–132

31. Z. Xu. (1997) Fuzzy logic control system CAD and study on the improvement of FLC performance. Master thesis, Guangxi University, Nanning, Guangxi P. R. China

32. Z. Xu. (2001) Fuzzy logic techniques for software reliability engineering. Ph. D. dissertation, Advised by Professor Taghi M. Khoshgoftaar, Florida Atlantic University, Boca Raton, Florida

33. J. Yen and R. Langari. (1999) *Fuzzy Logic: Intelligence, Control, and Information*. Prentice Hall, Inc., Upper Saddle River, New Jersey

11 Processing Software Engineering Data: Granular-based Approach

Marek Reformat[1] and Witold Pedrycz[1,2]

[1] University of Alberta, Edmonton, Canada,
[2] Systems Research Institute, Polish Academy of Sciences 01-447 Warsaw, Poland

11.1 Introduction

There are many definitions of Software Engineering. The one by IEEE states that Software Engineering is "the application of a systematic, disciplined, quantifiable approach to the development, operation, and maintenance of software; that is, the application of engineering to software" [1]. Despite a variety of definitions, there is a common denominator in all of them - indication of methodologies and techniques supporting the construction, deployment and evolution of software. In order to develop effective and useful methodologies, techniques and tools, a solid understanding of software development processes is needed. Deep and thorough knowledge about rules and mechanizes that govern software development is essential. Software differs from most products. It is developed rather than produced, and has a nonvisible nature - it is difficult to see the structure or the function of software. This very nature of software means that special techniques should be used to learn about software and processes of software development. To ensure a successful development, software engineers and developers shall understand all components, relations, rules and constrains related to software development. An interesting question is what techniques and methods should be used to extract such knowledge and gain understanding of software processes. One of the possible approaches is analysis of software data.

Analysis of software data, similar to data mining [5,7], can be defined as the process of inducing previously unknown, and potentially useful, information from software engineering data sets. Major areas of data processing include

- building of models, for example models for fault prediction of software modules where the mined information is contained in the derived model itself;
- automatic extraction of patterns, for example patterns containing associations between software project characteristics and software faults profiles, in this case the mined information will be extracted when a domain expert examines the patterns;
- visual exploration of data, for example usage of interactive data visualization tools to obtain high level views of the projective characteristics

and fault profiles, as in the previous case the mined information will be extracted by a domain expert.

In general, data processing tasks can be roughly classified into six categories: estimation and prediction, classification, association discovery, clustering, visualization of data and visual data exploration. The first two of them, classification and prediction, aim to build explicit models, which are ready to be employed in the software organization. The next two tasks, association discovery and clustering, aim to automatically identify useful patterns in data. Those patterns have to be interpreted by a domain expert in order to produce business insights for an organization. The last two techniques, visualization and visual data exploration, aim to help domain experts to easily find, by themselves, useful patterns in the explored data. These patterns can be used to produce business insights or to simply help the experts to better understand what is going on in the data.

All data processing activities are performed on software engineering data, which represents and describes software products and processes. A format of this data should be such that automatic or semi-automatic processing of data is possible. This requirement is fulfilled when software is represented by software metrics. This approach has already been identified as one of the ways of dealing with qualitative problems in a quantitative way [3,4]. Software metrics can be extracted directly from software. In this case they illustrate different features and attributes of software components. Measures can also identify features which are not directly extracted from software and are related to such processes as construction, testing, maintenance and even evolution of software components. In this case measures can represent such attributes of software components as number of defects or component readability. Data processing activity presented in the paper combines two distinctive tasks: prediction and association discovery. In the area of prediction, the proposed approach builds a model by examining attributes of a set of software components and, based on these attribute values, assigning values to an unknown attribute that one wants to quantify. At the same time the model is being developed, an association discovery takes place. An identification of attributes that are associated with each other is performed and the relationships between these attributes are determined. Fusion of these two tasks is accomplished by synthesis of models capable of dealing with two highly conflicting design requirements such as accuracy of a model and transparency. The proposed process of software model development is base on the following concepts

- granular computation - granules of software measures are used in analysis of software processes, the resulting granular-based models exhibit a high level of comprehension of processed data;
- logic-based modeling - models with well-defined logic fabric become a prerequisite to models' transparency and user-friendliness;

- multiphase model development process - clear separation of structural and parametric development, the process starts with building a "skeleton" (more qualitative than quantitative) or a blueprint of the model that is concerned with the structural relationships, then its further numeric refinement (parametric optimization) is performed.

The proposed approach is used to design a model representing a process of estimation of development efforts of software modules.

The paper is organized into 5 sections. First, in Section 11.2 basic concepts of granulation and granular-based models are introduced. Section 11.3 describes evolutionary-based technique used for synthesis of model. Section 11.4 covers details of data analysis and an overall modeling of a process of estimating maintenance efforts of software modules. Section 11.5 covers concluding remarks.

11.2   Granular Computing

11.2.1   Granularity

Versatility and complexity of software processes and products, involvement of human beings at all stages of software development and difficulties of software measurement processes make software data difficult to analyze. In such case, extraction of valuable, useful information from software data is extremely tiresome. In order to concentrate on essential issues "hidden" in the data, as well as isolate details and imperfections, an approach is proposed to enhance analysis of software data by the technology of granular computing.

Granular computing is geared toward representing and processing information granules. Information granules are collections of entities, usually originating at the numeric level, that are arranged together due to their similarity, functional adjacency, indistinguishability or alike [13]. This approach is suitable in situations when there is a need to comprehend the problem and provide a better insight into its essence rather than get buried in all unnecessary details. In this sense, granulation serves as an important mechanism of abstraction. This approach allows hiding some flaws in an analyzed data making the analysis more immune to outliers.

Granular computing is concerned with a representation of knowledge in terms of information granules. In particular, fuzzy sets can be considered as one of the possibilities existing there. This option is of interest owing to the clearly defined semantics of fuzzy sets so that they can formalize the linguistic terms being used by users/designers when analyzing data. One can easily change the size of information granules (by adjusting the membership functions) so that the resulting data analysis can be completed concentrating on a certain level of details to be taken into consideration. It has to be noted that the last step of the data processing is to assimilate the extracted information. This is usually done by a domain expert who will look at the mined information and use his background knowledge to check if this information is indeed something new, useful, and non-trivial (i.e., new knowledge). Usage of granular computing has a big advantage for an expert. Granular-based models are user-friendly and the rules that govern relationships between attributes can be extracted. The domain expert can easily sanity check the model. Moreover, he/she may use the information explicitly described in the models to gain new knowledge about he domain being analyzed. In other words, a proposed model building tool sheds some light on the problem at hand.

11.2.2   Granular-based Modeling

Representation of data using information granules leads to application of models, which have abilities to deal with granules and express relations between them. Fuzzy models which hinge on information granules are such models with fuzzy sets seen as information granules. The interaction of the models with the environment, which is predominantly numeric, is realized through model interfaces. The input interface - encoder or fuzzification block - transforms input data into the format acceptable by fuzzy set granules. The output interface - decoder or defuzzification module - translates the results from the level of information granules into the numeric format of the modeling environment, where the original numeric data come from. As discussed in [14], two main modeling and ensuing development scenarios can be envisioned, Fig. 11.1. The first one is the external optimization of the fuzzy model performed at the level of numeric experimental data. From the functional standpoint, the fuzzy model is a structure of three mappings put in series, that is the encoder $\mathbf{x} = \text{ENC}(x)$, processing part of the fuzzy model $\mathbf{y} = FM(\mathbf{x}, \mathbf{p})$, and the decoding part $y = \text{DEC}(\mathbf{y})$ where $\mathbf{x}$ and $\mathbf{y}$ are elements in the domain of information granules, while x and y are input-outputs of the model at the numeric level. The experimental data are given in the form of input-output numeric pairs, say $\{x(k), target(k)\}$, where k=1, 2, ..., N. The requirement is that the output of the fuzzy model $y(k)$ for the input $x(k)$ is equal to $target(k)$, it means $y(k) \approx target(k)$. The optimization of the parameters of the fuzzy model ($\mathbf{p}$) requires a backpropagation of the error through the decoder. In other words, the gradient - based scheme requires the calculations of the following expression

$$\frac{\partial y(k)}{\partial p_i} = \frac{\partial DEC(\mathbf{y})}{\partial FM(\mathbf{x}, \mathbf{p})} \frac{\partial FM(\mathbf{x}, \mathbf{p})}{\partial p_i}$$

where "i" denotes an i-th position of the vector of the parameters of the fuzzy model. It becomes evident that the form of the decoder plays an important role in the optimization process because of the chaining effect i.e., a way in which the parameters of the model ($\mathbf{p}$) can be "accessed" for optimization purposes. Furthermore as both the encoder and decoder are included in the model identification, they may be optimized as well.

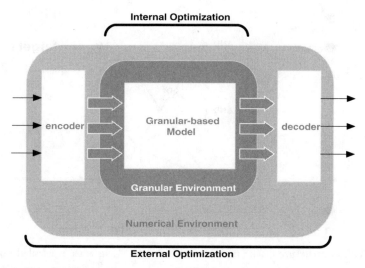

Fig. 11.1. Two fundamental scenarios of the development of fuzzy models: external optimization at the level of numeric data, and internal optimization at the level of fuzzy sets in the input and output space

While the first mode could be viewed an external mode of fuzzy identification, the second one is an internal one. In this case, the processing module is optimized. The original training data $\{x(k), target(k)\}$ are granulated through fuzzy sets occurring in the encoding and decoding blocks. These fuzzy sets are considered to be fixed in advance. As the result the experimental data is in the format of input-output pairs $\{\mathbf{x}(k), \mathbf{target}(k)\}$ and as such used for the optimization of the fuzzy model. In other words, a requirement $\mathbf{y}(k) \approx \mathbf{target}(k)$ k=1, 2, ..., N is used as optimization goal. More realistically, a sum of squared errors needs to be minimized with regard to the structure of the fuzzy model (FM) and its parameters,

$$Q = \sum_{k=1}^{N} (FM(\mathbf{x}(k) - \mathbf{target}(k))^{T} (FM(\mathbf{x}(k) - \mathbf{target}(k)) \Rightarrow Min$$

In this study, the main emphasis is put on the granular modeling which it is geared toward internal optimization. This type of optimization promotes the interpretability of the fuzzy model due to the focus on the mapping of the fuzzy sets rather than on the numeric experimental data.

The structure of the fuzzy neural network is fully determined by the network. The learning of the network leads to its further refinements that appear at the numeric level. Proceeding with the architectural details, the fuzzy neural network is governed by the following expressions, refer also to Fig. 11.2.

The mapping from the structure to the fuzzy neural network is straightforward. Recall that an n-input single output OR neuron is described in the

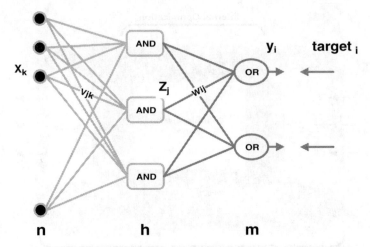

Fig. 11.2. A structure of the fuzzy neural network along with a detailed notation

form

$$y = OR(\mathbf{z};\ \mathbf{w})$$

where $\mathbf{z}, y \in [0, 1]$. The connections $w_1, w_2, ..., w_n$ are arranged in a vector form ($\mathbf{w}$). Rewriting the above expression in a coordinate wise manner results in

$$y = \mathbf{S}_{i=1}^{n}(z_i\ t\ w_i)$$

meaning that the neuron realizes an s-t composition of the corresponding finite sets $\mathbf{z}$ and $\mathbf{w}$.

The AND neuron $z = AND(\mathbf{x};\ \mathbf{v})$ is governed by the expression

$$z = \mathbf{T}_{i=1}^{n}(x_i\ s\ v_i)$$

Computationally, this neuron realizes a standard t-s composition of $\mathbf{x}$ and $\mathbf{v}$.

The role of the connections in both neurons is to calibrate the inputs and in this way furnish them with the required parametric flexibility. In case of OR neurons, the higher the connection, the more essential the associated input. For AND neurons an opposite situation holds: lower connection indicates that the respective input is more essential. In general, a certain thresholding operation can be sought. For any OR neuron, the input is irrelevant if the associated connection assumes values lower than 0.5. An input of the AND neuron is viewed irrelevant if the connection exceeds 0.5. A fuzzy model is built in a process of parametric learning using a gradient-based optimization scheme [12]. The details of the on-line learning algorithm are given below

connections are updated after processing

of each input-output pair of data $\{\mathbf{x}, \mathbf{target}\}$

$$w_{ij}(iter+) = w_{ij}(iter) - \alpha \frac{\partial Q}{\partial w_{ij}(iter)}$$

$$v_{ij}(iter+) = v_{ij}(iter) - \alpha \frac{\partial Q}{\partial v_{ij}(iter)}$$

where
$$Q = (\mathbf{target} - \mathbf{y})^T (\mathbf{target} - \mathbf{y})$$
$w(0)$, $v(0)$ have randomly generated values

The computations carried out there deal with the triangular norm and co-norm specified as a product operation and a probabilistic sum. The general learning scheme can be made more specific once particular triangular norms are selected.

The parametric learning of the fuzzy neural network has been well developed and documented in the literature [14,12]. Several general observations are worth summarizing

- The gradient-based learning supports optimization that may result in a local minimum of the performance index. Global minimum could be out of reach of this learning mechanisms.
- The efficiency of learning depends upon the choice of the triangular norms and co-norms. Here the minimum and maximum operators deserve particular attention as they lead to optimization traps. One of the reasons is that both minimum and maximum are non-interactive meaning that the results depends on an extreme value encountered there and the final outcome does not reflect the remaining arguments of these t- and s-norms. The other hand, for most other t-norms a saturation effect can occur that may be extremely difficult to handle in case of higher dimensions of the problem. For instance, consider the product as a model of the t-norm. If the number of arguments increases, the result of aggregation carried out in this way tends to zero. Now if one envisions such an AND neuron located in the input layer of the fuzzy neural network and assume that all connections are the same and equal to zero, the output of the neuron reads as $z = \times_{i=1}^{n} x_i$. For any input less than one, say $1 - \gamma$, the output equals to $(1 - \gamma)^n$. One can easily check that a situation in which $\gamma = 0.5$ and $n = 40$ inputs produces the output of the neuron equal to $9.095 * 10^{-13}$. This activation level reduces quickly once the dimensionality of the problem goes up.
- The learning may be very slow especially when the size of the network gets large. Initialization of the connections with random values and no preliminary knowledge about the structure of the network (that implies its fully connected topology where all neurons are connected with the

neurons in the neighboring layer) means exposure to the curse of dimensionality.

In light of these observations, the general design paradigm proposed in this study is strongly supported. Instead of learning the fuzzy neural network from scratch (the process which may fail quite easily), a structural blueprint of the network is established first and then the learning of the connections continuous. Effectively, this skeleton of the network reduces the number of connections to be learned. The structural optimization of the network is out of reach of parametric (gradient-based) optimization and requires methods along the line of Evolutionary Computing [10,16].

11.3   Evolutionary-based Synthesis of Granular Models

11.3.1   Concept

The evident diversity of fuzzy models existing in the literature is amazingly uniform in terms of their underlying design. In a nutshell, the original numeric data are transformed through fuzzy sets and this is followed by the construction (estimation) of the parameters of the fuzzy model. In contrast, the proposed development of the fuzzy model is comprised of the three fundamental phases

- A collection of fuzzy sets (or fuzzy relations) is established. They are viewed as basic conceptual "building" blocks of the fuzzy model. Intention is to regard these entities to be as independent as possible from the ensuing detailed structure of the model. In this sense, these information granules stand on their own and exhibit a well-defined semantics.
- A structure of the fuzzy model is looked for. Evidently, any structural optimization of the model is far more challenging than a parametric optimization. At this phase, evolutionary computing is exploited to optimize the structure. Furthermore the structural optimization is carried out independently from parametric optimization. By distinguishing between the structure and the parameters the emphasis is put on the topology of the model. This approach makes structural optimization disjoint from the phase concentrated on parameter adjustment. In this case, the structure is searched for by exploiting the space of all possible structures. When the best structure is found, the process of fine tuning is performed at the parametric level. There is a crucial reason behind the use of evolutionary techniques. First, the structural optimization is not supported by gradient-based techniques. Second, the space of the structures is large and this calls for the use of evolutionary techniques. As to the structure itself, a standard two-level OR-AND representation of functions of symbols (in this phase fuzzy sets are used as symbols) is used. Interestingly, this representation is in line with the well-known structures of rules (if-then statements) composed of fuzzy sets standing in their condition and conclusion parts.

- Once the topology of the network has been established during the previous phase, the network is subject to some parametric refinement through learning of the induced fuzzy neural network.

11.3.2   Genetic Algorithms

Genetic Algorithms (GA) [8] are one of the most popular methods of Evolutionary Computation. They have been successfully applied to numerous problems [2,6,8] both at the level of structural and parametric optimization.

GA is a search method utilizing the principles of natural selection and genetics [9]. In general, GA operate on a set of potential solutions to a given problem. These solutions are encoded into chromosome-like data structures named genotypes. A set of genotypes is called a population. The genotypes are evaluated based on their ability to solve the problem represented by a fitness function. This function embraces all requirements imposed on the solution of the problem. The results of the evaluation are used in a process of forming a new set of potential solutions. The choice of individuals to be reproduced into the next population is performed in a process called selection. This process is based on the fitness values assigned to each genotype. The selection can be performed in numerous ways. One of them is called stochastic sampling with replacement. In this method the entire population is mapped onto a roulette wheel where each individual is represented by the area corresponding to its fitness. Individuals of an intermediate population are chosen by repetitive spinning of a roulette wheel. Finally, the operations of crossover and mutation [15] are performed on individuals from intermediate population. This process leads to the creation of the next population. Crossover allows an exchange of information among individual genotypes in the population and provides innovative capability to the GA. It is applied to the randomly paired genotypes from the intermediate population with the probability p_c. The genotypes of each pair are split into two parts at the crossover point generated at random, and then recombined into a new pair of genotypes. Mutation, on the other hand, ensures the diversity needed in the evolution. In this case all bits in all substrings of genotypes are altered with the probability p_m.

All steps described above are then repeated. Every iteration is referred to as a generation. A pseudo-code of Genetic Algorithm is presented below

```
generation_number := 0
creation_of_initial_population()
evaluation_of_population ();
while not terminate do
    selection()
    crossover ()
    mutation ()
    generation_number++
```

evaluation_of_population()

od

11.3.3   Genetic Algorithms-based Construction of Granular-based Models

Having identified the shortcomings of the learning in the fuzzy model, an intuitively appealing approach has been proposed. It relays on a hybrid learning methodology comprising of two fundamental techniques, namely genetic optimization and gradient-based optimization. To battle the dimensionality problem, a skeleton of the network is constructed by detecting the most essential connections that shape up the architecture and then the detailed optimization of these connections is performed. As the blueprint, the network has to capture the essence of the mapping of an input into an output. Therefore the connections are binary (0-1) for the purpose of this optimization. A standard version of a Genetic Algorithm [8,9] is applied. A binary string represents the structure of the network. As the connections are two-valued, the entries of the string are mapped directly on the connections of the neurons, Fig. 11.3. The treatment of the connections as binary entities is highly desirable. By doing this, the effort is put on the determination of the most essential structure of the model that attempts to reveal the crux of its topology.

The binary connections of the neuron have a straightforward interpretation. In case of the AND neuron: If the connection is equal to zero, this means that the corresponding input impacts the output of this neuron. For the OR neuron a reverse situation occurs: the connection equal to one implies that the specific input is essential and affects the output of the neuron. From the design standpoint, the development of the topology of the network can be easily controlled. In particular, the level of connectivity between the input and hidden layer can be fixed. A maximal number of connections from the input nodes to each node (AND neuron) in the hidden layer is referred to as the network connectivity. For instance, having two inputs to the first AND neuron, four to the second and three to the third, it can be said that the connectivity here is equal four. Such connectivity analysis helps understand the nature of data. First, the number of the connections of the AND neuron determines a local dimensionality of the input space. The contributing inputs form this local space. By varying the maximal number of the connections that are allowed, one can gain a better insight into the intrinsic dimensionality of the data space. The connections of the OR neuron indicate which of these local input spaces contribute to the output of the network.

An initial configuration of the skeleton of the network is set up as follows. For the given number of the neurons in the hidden layer and the connectivity level (that is the number of the inputs going to each neuron in the hidden layer), the connections between the AND neurons and inputs are randomly initiated (making sure that the number of the connections between the inputs and each neuron is as specified). In same is in the case of the connections of

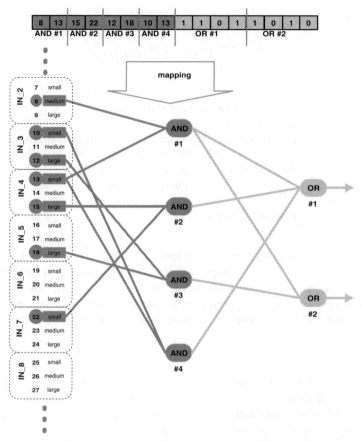

Fig. 11.3. A part of the network and its one-to-one correspondence with the binary string being a chromosome of GA. The details of the structure of the network pertain to the data with 36 inputs (12 input variables with 3 fuzzy sets per variable) 2 outputs (that is 2 fuzzy sets defined for the output variable) and connectivity of 2 inputs per node in the hidden layer, and 4 nodes in the hidden layer

the OR neurons. Then the genetic optimization adjusts the connections. The adjustment of connections is driven by fitness function, which is an objective of an optimization process. Details of these elements are as follow:

- Fitness function
  The role of the fitness function is to assess how well the model matches the experimental data. Fuzzy sets are defined in the input spaces. While they show up in the lists as symbols (say, A1, B3, etc.), their membership functions are present in the computations of the fitness function. The fitness function is computed in the form:

$$Fit\_Fun = \frac{1}{1 + Q}$$

where Q is a commonly used performance index assuming the form

$$Q = \sum_{k=1}^{N} (\mathbf{F}(k) - \hat{\mathbf{F}})^T (\mathbf{F}(k) - \hat{\mathbf{F}})$$

with:

N - number of training data points

m - number of outputs

$F_i(k)$ - value of the output "i" for given training
data point "k" obtained from the model

$\hat{F}_i(k)$ - original value of the output "i"
for given training data point "k"

moreover

$$\mathbf{F}(k) = [F_1(k), F_2(k), ..., F_m(k)]^T$$

- Crossover
  The role of this operator is to explore the search space formed by a collection of lists. The mechanism of a possible crossover is presented in Fig. 11.4. An arbitrary point in parent strings is picked and the information is exchanged between parents. As the result two new strings are obtained which have beginnings and ends coming from different parent strings.

- Mutation
  The role of mutation is to introduce diversity into a population and promote a probabilistic-like traversing of the search space. As shown in Fig. 11.5, mutations form a new binary string by making an alteration to the string in a copy of a single, parent string.

11.4   Module Development Efforts

Software development processes include design, construction, testing and different types of maintenance activities. In this case of maintenance, for example, it is important to know relationships between construction of software components and a number of corrections and modifications, which these components have to go through. A model is constructed which expresses relationships between software measures representing software components and a number of needed modifications of each component. knowledge about these relationships can lead to increased accuracy of estimates of software development processes, mostly the phases of testing and maintenance.

11.4.1   Description of Data set

The Medical Imaging System (MIS) is a commercial software system consisting of approximately 4500 routines. It is written in Pascal, FORTRAN and

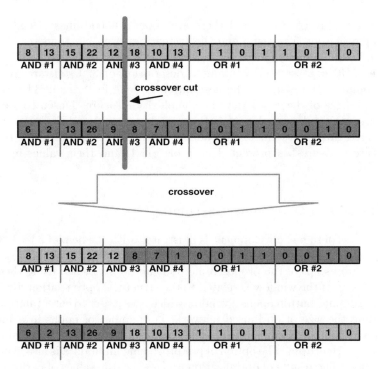

Fig. 11.4. Crossover operation used in genetic optimization of the model

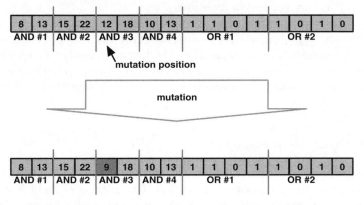

Fig. 11.5. Mutation operation used in genetic optimization of the model

PL/M assembly code. In total there are about 400,000 lines of code. MIS development took five years, and the system has been in commercial use at several hundred sites.

The MIS data set [11] contains a description of 390 software modules that comprise the system. The description is done in terms of 11 software measures (size of the code, McCabe complexity measure, Jensen complexity measure, etc) and a number of changes made to each module due to faults discovered during system testing and maintenance. A list of all attributes (software measures) is presented in Table 11.1, and their values in Table 11.2.

11.4.2   Model Development System

The process of model development is illustrated with the help of a tool "FNN Optimizer". A dialog box of it is shown in Fig. 11.6. Structural and parametric optimization steps of the proposed method can be performed using the tool. The left side of the window is related to the structural optimization. Pressing the "GA Input" button opens a window which asks a user to enter information regarding the size of optimized network, i.e. number of nodes in a hidden layer, and number of connectivity. The user also enters values of parameters of Genetic Algorithm. The option of pre-processing input data is also available to the user. The results of optimization process, i.e. the values of performance index and indications which connection are selected, are shown on the left side of application window in real-time, Fig. 11.6.

Once the structural optimization of a model is finished the user can start its parametric optimization. In this case, the "Gradient Input" button invokes a small windows which prompts the user to enter number of epochs and learning rate. The values of performance index are shown to the user together with the values of weights of connection chosen during structural optimization process.

11.4.3   Analysis of results

The fuzzy models are developed based on 2/3 data points form the MIS data, with the rest being used as a test set. Granulation of input is done by defining 3 fuzzy sets in each input space (attribute) with linguistic labels: *small, medium,* and *large.* In the output space, which represents a number of changes in a module, there are only 2 granules defined: *small* and *large.* This means that a developed model has two outputs. Each of the outputs represents a single fuzzy set (granule). All fuzzy sets are described by means of Gaussian membership functions.

A pre-processing of MIS data is performed first. Fig. 11.7 shows a dialog box of the "FNN Optimizer". A user has to enter a number of membership functions in the input and output spaces, as well as the level at which all

Table 11.1. List of attributes of the MIS data

LOC	length of code with comments
CL	length of code without comments
TCHAR	number of characters
TCOMM	number of comments
MCHAR	number of comment characters
DCHAR	number of code characters
N	program length, $N = N_1 + N_2$, where N_1 is the total number of operators, and N_2 is the total number of operands
NE	estimated program length, $NE = \eta_1 \, log_2 \eta_1 + \eta_2 \, log_2 \eta_2$ where η_1 is the number of unique operators and η_2 is the number of unique operands
NF	Jensen's program length, $NF = (log_2 \eta_1)! + (log_2 \eta_2)$
VG	McCabe's cyclomatic number, is one more than the number of decision nodes in the control flow graph
BW	Belady's bandwidth metric, where $BW = \frac{1}{n} \sum i L_i$ where L_i represents the number of nodes at level i in a nested control flow graph of n nodes, this metric is indicative of the average level of nesting or width of the control flow graph representation of the program
NC	number of changes

membership functions cross with each other (all functions are distributed uniformly).

A series of experiments for two parameters controlling the structure of the network, that is a level of connectivity and the number of AND nodes in the hidden layer, has been carried out. The results for the training set are presented in Table 11.3. The performance index reported there is taken as an average error per data point. Its value for the optimal structure is equal to 0.0219. While using a testing dataset, the value of the normalized performance

Table 11.2. Set of software measures

Feature	Min	Max	Mean
LOC	0	5669	296.7
CL	0	108	15.9
TCHAR	0	26611	1387.2
TCOMM	0	16383	1051.2
MCHAR	0	582	25.1
DCHAR	0	21	1.6
N	0	1061	51.6
NE	0	91	3.4
NF	0	20	0.7
VG	0	3	0.4
BW	0	10	0.3
NC	0	16	1.0

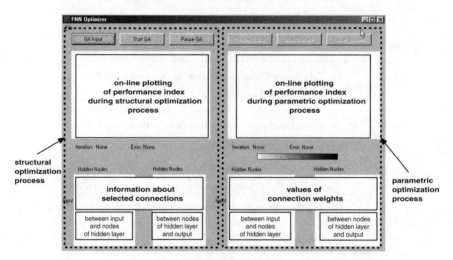

Fig. 11.6. FNN Optimizer

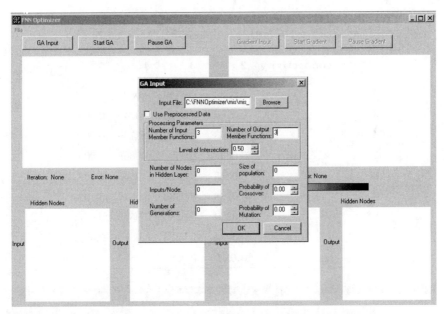

Fig. 11.7. FNN Optimizer with a dialog box for pre-processing of the data

index rises a bit and equals to 0.0290. The optimal structure of the model contains 10 nodes in the hidden layer, with 2 inputs to each of these nodes. It can be said that for a fixed connectivity, the increase of the size of the hidden layer results, in general, in lower values of the performance index. On the other hand, it is interesting to learn that the increased connectivity for a fixed size of the hidden layer gives rise to the higher values of the performance index.

FNN Optimizer is used to illustrate the structural optimization phase of a model building process. The values of GA parameters are entered into a dialog box, Fig. 11.8.

The results of the structural optimization process are shown in Fig. 11.9. In the upper left corner of the application window the changes in the values of performance index during the optimization process can be observed. In the lower left corner of the box a simplified "image" of the model is presented. In the first box, with "Input" on y axis and "Hidden Nodes" on x axis, one can see choices of inputs for each hidden layer node - black bars indicate a choice. It can be easily seen that each hidden layer node has two input – there are two black bars in each column. The second box, with "Output" on y axis and "Hidden Nodes" on x axis, represent chosen connection between hidden layer nodes and output nodes. In the case shown, there are two outputs: the upper row represents all the connections between hidden layer nodes and the output node corresponding to the output *small*, the lower row shows the

Table 11.3. The results of the GA optimization of the network for the training set (case in the boldface is used in further optimization of the model)

hidden/connectivity	2	3	4	5
2	0.0330	0.0546	0.0825	0.1127
3	0.0265	0.0428	0.0745	0.1060
4	0.0253	0.0419	0.0706	0.1079
5	0.0245	0.0374	0.0668	0.0966
6	0.0238	0.0312	0.0648	0.1012
7	0.0228	0.0330	0.0542	0.1032
8	0.0229	0.0286	0.0543	0.0971
9	0.0224	0.0278	0.0540	0.0997
10	**0.0219**	0.0294	0.0509	0.0977

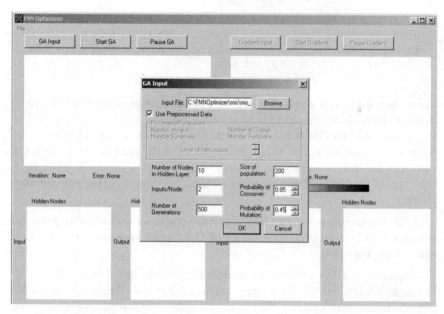

Fig. 11.8. Preparation stage for GA structural optimization of a model

connection between hidden layer and the output *large*. The architecture of the constructed network is presented in Fig. 11.10.

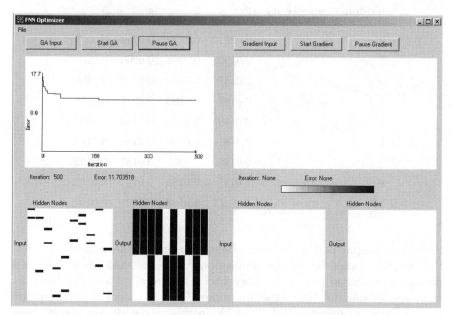

Fig. 11.9. Performance index Q in successive generations of GA optimization, together with a graphical representation of the obtained model

In the next development phase a blueprint of the network, produced by GA, is an object of parametric optimization. The gradient-based learning starts with connections established by GA optimization and transforms them into the weights. This enhancement aims at further reduction in the value of performance index, Table 11.4.

The normalized performance index of the optimal fuzzy neural network, after structural and parametric optimization, is equal to 0.0177, and for the testing set becomes equal to 0.0230. The value of the index, comparing with its initial value after the structural optimization, improves by 19% in the case of training data, and by 21% in the case of testing data. The optimal model has the structure that contains 10 nodes in the hidden layer and 2 inputs to each of these nodes.

Again, illustration of the parametric optimization using FNN Optimizer is shown in Fig. 11.11 and Fig. 11.12. Fig. 11.11 presents initial phase of parametric optimization, and Fig. 11.12 shows the values of performance index versus successive learning epochs.

The window of "FNN Optimizer", the lower right corner of it - Fig. 11.12, contains also manifestation of modified weights of the connections between inputs and nodes in hidden layer, and the connections between hidden layer

Table 11.4. The results of the parametric optimization of the network for the training set)

hidden/connectivity	2	3	4	5
2	0.0201	0.0199	0.0204	0.0205
3	0.0190	0.0204	0.0200	0.0192
4	0.0187	0.0199	0.0192	0.0190
5	0.0191	0.0193	0.0217	0.0259
6	0.0190	0.0188	0.0194	0.0207
7	0.0181	0.0188	0.0188	0.0193
8	0.0178	0.0192	0.0185	0.0191
9	0.0185	0.0186	0.0194	0.0192
10	**0.0177**	0.0183	0.0191	0.0197

nodes and output nodes. One can observe changes in the values of connection weights comparing to the connections of the network after structural optimization. Some connection have been eliminated, other ones have become less meaningful. The structure of the MIS data model after parametric optimization is presented in Fig. 11.13. In this model all connections have been augmented by the values, which change the significance of some input sets and the rules. Only valid connections (with weights below 0.5 for AND nodes, and above 0.5 for OR nodes) are presented. Further simplification of a network can be observed.

If-conditions of the network, which represent the relationships between the software measures describing a software module and the number of modification of the module, are shown below

if
    LENGTH_OF_CODE_WITH_COMMENTS is small (LOC_small)
or
    BELADYS_BANDWIDTH_METRIC is medium (BW_medium)
and
    ESTIMATED_PROGRAM_LENGTH is small (NE_small)
or
    NUMBER_OF_COMMENTS is large (TCOMM_large)
and
    LENGTH_OF_CODE_NO_COMMENTS is medium (CL_medium)
or
    LENGTH_OF_CODE_WITH_COMMENTS is small (LOC_small)
and
    PROGRAM_LENGTH is large (N_large)
or

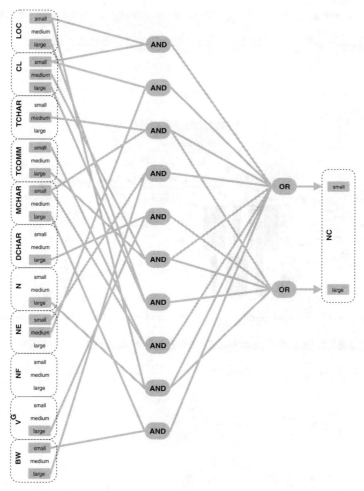

Fig. 11.10. A network generated by GA optimization process

    NUMBER_OF_COMMENTS is small (TCOMM_small)
then
    NUMBER_OF_CHANGES is small (NC_small)
if
    LENGTH_OF_CODE_WITH_COMMENTS is large (LOC_large)
and
    LENGTH_OF_CODE_NO_COMMENTS is medium (CL_large)
or
    NUMBER_OF_CODE_CHARACTERS is large (DCHAR_large)
and
    MCCABE_CYCLOMATIC_NUMBER is medium (VG_large)
or

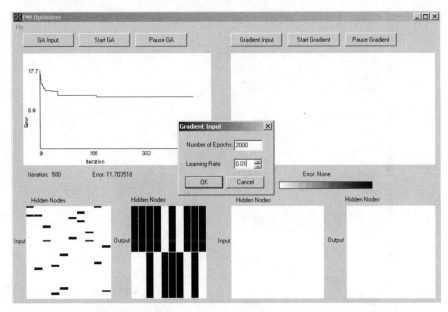

Fig. 11.11. Process of entering the parameters of gradient-based learning

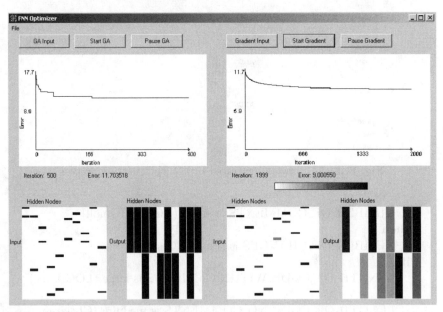

Fig. 11.12. Performance index Q in successive learning epochs for parametric learning of fuzzy neural network representing the MIS data

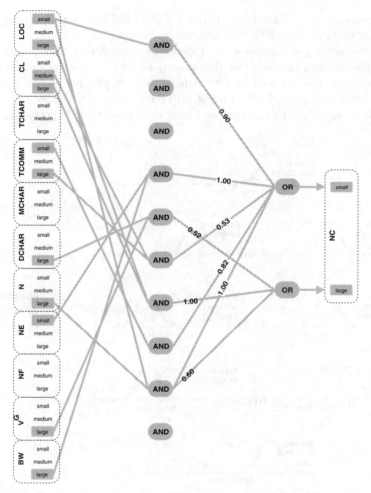

Fig. 11.13. A network generated by GA optimization process and tuned by parametric optimization; shown are only valid connections

> LENGTH_OF_CODE_WITH_COMMENTS is small (LOC_small)
> and
> PROGRAM_LENGTH is large (N_large)
> **then**
> NUMBER_OF_CHANGES is large (NC_large)

Graphical representations of the rules, with granules defined in the domain of each measure, are shown in Fig. 11.14 and Fig. 11.15 for the cases of small and large NUMBER_OF_CHANGES respectively. One can easily notice intuitive character the rules. Rules representing a *small* NUMBER_OF_CHANGES contain a significant number of granules with linguistic labels *small* and *medium* distributed among six different software measures. In the case of rules representing a *large* NUMBER_OF_CHANGES most granules are with a linguistic label *large* form five software measures.

These rules give managers and developers indication about quality of development components. Once they determine measures describing a module they can use the rules to better estimate testing and maintenance efforts related to the module. In the case of MIS data, some of the rules seems easy to induce. However, this can be said after data processing is performed.

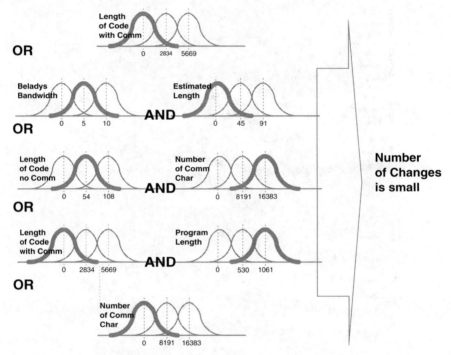

Fig. 11.14. Rules for NUMBER_OF_CHANGES small

11.5 Conclusions

In this study, we have proposed a general design methodology for granular-based models fulfilling the two fundamental requirements of granular modeling that is accuracy and transparency. The optimization tandem of evolutionary computing and gradient-based learning of fuzzy neural networks naturally supports structural and parametric optimization of the models that helps achieve accuracy. The transparency of the model is accomplished by subscribing to the logic-oriented architecture of the fuzzy neural networks.

The proposed approach has been used to perform two tasks of processing software engineering data - prediction and association discovery. A model of a

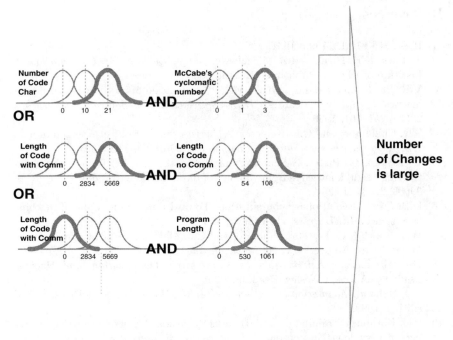

Fig. 11.15. Rules for NUMBER_OF_CHANGES large

process of development effort of software modules has been built. The model has led to the recognition of significant relationships between software measures. These relationships are essential for development purposes. Software mangers can use the model in two different ways. They can use it to predict a number of needed changes in a module during development process. They can also improve their knowledge about relationship between the values of a selected set of software measures and development effort.

One of the open issues worth pursuing is related to the matter of the input and output interfaces. They have been left out from the study. The careful design of the interface can lead to further enhancement of the performance of the fuzzy model. The related question is that of simultaneous of separate optimization of the input and output interfaces.

11.6   Acknowledgments

Support from the Natural Sciences and Engineering Research Council (NSERC) and Alberta Software Engineering Research Consortium (ASERC) is gratefully acknowledged.

References

1. IEEE Std 610.12, 1990, 1991.
2. T. Back, D.B. Fogel, and Z. Michalewicz (eds.). *Evolutionary Computations I.* Institute of Physics Publishing, Bristol, 2000.
3. V.R. Basili, L.C. Briand, and W. Melo. A validation of object-oriented design metrics as quality indicators. *IEEE Transactions on Software Engineering*, 22(10):751–761, 1996.
4. S.R. Chidamber and C.F. Kemerer. A metrics suite for object-oriented design. *IEEE Transactions on Software Engineering*, 20(6):476–493, 1994.
5. U. Fayyad, G. Piatetsky-Shapiro, and P. Smyth. The KDD process for extracting useful knowledge from volumes of data. *Communications of the ACM*, 39(11):27–34, 1996.
6. D.B. Fogel. Evolutionary computation. Toward a new philosophy of machine intelligence. *IEEE Press, Piscataway*, 1995.
7. W.J. Frawley, G. Piatetsky-Shapiro, and C.J. Mattheus. Knowledge discovery in databases: An overview. *AI Magazine*, pages 57–70, Fall 1992.
8. D.E. Goldberg. *Genetic Algorithms in Search, Optimization, and Machine Learning.* Addison-Wesley, Reading, 1989.
9. J.H. Holland. *Adaptation in Natural and Artificial Systems.* MIT Press, second edition, 1992.
10. P.G. Korning. "Training neural networks by means of genetic algorithms working on very long chromosomes". *International Journal of Neural Systems*, 6(3):299–316, 1995.
11. J.C. Munson and T.M. Khoshgoftaar. Software metrics for reliability assessment. In M.R. Lyu, editor, *Software Reliability Engineering*, pages 493–529. Computer Society Press, Los Alamitos, 1996.
12. W. Pedrycz. *Fuzzy Sets Engineering.* CRC Press, Boca Raton, FL, 1995.
13. W. Pedrycz. Granular computing: An introduction. In *Proceedings of Joint 9th IFSA World Congress and 20th NAFIPS International Conference*, pages 1349–1354, Vancouver, Canada, July 25-28 2001.
14. W. Pedrycz and F. Gomide. *An Introduction to Fuzzy Sets: Analysis and Design.* MIT Press, 1998.
15. W.M. Spears. "Crossover or Mutation?". In *Proceedings of Foundations of Genetic Algorithms Workshop*, pages 221–237, 1992.
16. X. Yao. "Evolving artificial neural networks". *Proceedings of IEEE*, 87(9):1423–1447, 1999.

Index